D1245035

Agritourism and Nature Tourism in California

Second Edition

Holly George

University of California Cooperative Extension

Livestock and Natural Resources Farm Advisor

Plumas-Sierra Counties

Ellie Rilla

University of California Cooperative Extension

Community Development Advisor

Marin County

Small Farm Center

University *of* **California**

Agriculture and Natural Resources

Publication 3484

To order or obtain ANR publications and other products, visit the ANR Communication Services online catalog at http://anrcatalog.ucdavis.edu or phone 1-800-994-8849. You can also place orders by mail or FAX, or request a printed catalog of our products from

University of California
Agriculture and Natural Resources
Communication Services
1301 S. 46th Street
Building 478 - MC 3580
Richmond, CA 94804-4600

Telephone 1-800-994-8849
510-665-2195
FAX 510-665-3427
E-mail: anrcatalog@ucdavis.edu

Publication 3484

ISBN-13: 978-1-60107-742-4
Library of Congress Control Number: 2011930690
Second edition, 2011

Photo credits are given in the Acknowledgments.

To simplify information, trade names of products have been used. No endorsement of named or illustrated products is intended, nor is criticism implied of similar products that are not mentioned or illustrated.

 This publication has been anonymously peer reviewed for technical accuracy by University of California scientists and other qualified professionals. This review process was managed by ANR Associate Editor for Farm Management and Economics Bees Butler.

 Printed in Canada on recycled paper

3m-pr-7/11-SB/CR

Contents

Acknowledgments

Our sincere thanks to the farmers and ranchers who shared their stories with us for this second edition of *Agritourism and Nature Tourism in California.* This book would not have been possible without the contributions from many talented individuals.

For writing and editing assistance of this second edition the authors thank Penny Leff, Jane Eckert, and the myriad operators who offered strategies and challenges. The contributors to the first edition included Sean Keith-Stewart, Roger Ingram, Robin Kozloff, and Marcie Rosenzweig, as well as Bob Barnes, Brian Crawford, Bob Garrison, John LeBlanc, Ramiro Lobo, Brett Matzke, Susan McCue, Didi Otterson, Dean Prigmore, Andrew Reid, Russel Reid, and Mary Wollenson.

For the many photographs, the authors thank Susan Giacomini Allan, USDA Agriculture Research Service; Jack Kelly Clark, University of California; Bob Garrison, OEM Defence Photographic Library; Patrick Laherty; Elizabeth Ptak; Leona Reed; Penny Leff; Roberta Almerez; Gary Romano; Jane Young; Tim Friesen; Roger Ingram; Feather River Land Trust; and Christian Ahlmann. In addition, the following USDA photographers deserve special thanks: Gordon Baer, William E. Carnahan, Irwin W. Cole, Ken Hammond, Russell Lee, Tim McCabe, Larry Rana, Arthur Rothstein, Irving Rusinow, Jack Schneider, Don C. Schuhart, Bill Tarpenning, John White, Doug Wilson, and Fred S. Witte.

For their reviews of early chapters, the authors acknowledge Bob Barnes, Brian Crawford, Bob Garrison, Eileen Hook, John LeBlanc, Ramiro Lobo, Brett Matzke, Susan McCue, Didi Otterson, Dean Prigmore, Andrew Reid, Russel Reid, Don Vetter, and Mary Wollenson.

In addition, the authors thank several state organizations and educational institutions for providing permission to adapt their resources; acknowledgment has been given at the end of applicable chapters.

The initial development of this book was funded by a grant from the Renewable Resources Extension Act and by funds from the University of California Agriculture and Natural Resources Ag/Nature Tourism Workgroup and the UC Small Farm Center.

Special Acknowledgment

The authors want to especially thank Diana Keith for her dedicated service as a technical writer for the initial version, produced in 2002, of what would become the first edition of *Agritourism and Nature Tourism in California*. We also want to recognize Desmond Jolly, former director of the UC Small Farm Center, who supported this idea and project from its beginning, providing Center funding and his expertise, guidance, and oversight.

We also thank Shermain Hardesty, current director of the UC Small Farm Program, for supporting agritourism research and the online farm and ranch database calagtour.org, and for providing Center funds to hire the first state agritourism coordinator.

Holly George, Ellie Rilla

About This Book

Since our first agritourism handbook was produced in 2002 and the first edition of *Agritourism and Nature Tourism in California* appeared in 2005, much has changed.

- Over 2,500 of you have attended workshops across the state.

- UC researchers completed the first-ever statewide survey of agritourism operators.

- Agritourism was officially recognized by the American Farm Bureau Federation.

- Several California counties amended or created ordinances to support and ease permitting this endeavor.

- The Small Farm Program at UC Davis created an online agritourism directory for California (http://calagtour.org) and expanded their listing of all the agritourism farms and ranches in the state.

- The first state agritourism coordinator was hired in 2009, proof that this type of business is growing.

Each chapter is updated with new information. Chapter 1 gives you the most recent data on consumer travel and buying preferences, with links to more information if you want to dig deeper. Chapter 2 has information about why other California operators started their agritourism businesses and who their customers are. Chapter 3 guides you through the process of developing your own business plan, with new links to funding sources. Chapter 4 shares recent changes to county, state, and federal policies that might be useful to you as you diversify your operations. Chapter 5 lists great publications now available about child and farm safety for agritourism operations. Chapter 6 describes the social media networking happening across the country with tips and ideas about how you can use this to your best advantage in marketing your farm; it also contains an extensive listing of all the marketing organizations in the state. Chapter 7 is completely revised with new agritourism-related contacts and links.

About Agritourism and Nature Tourism

Tourism is the largest industry in the world. Agricultural tourism—"agritourism"—and nature tourism are especially popular. Agritourism consists of businesses conducted by farmers or ranchers on their working agricultural, horticultural, or agribusiness operations for the enjoyment and education of visitors. It is intended to promote farm products and to generate additional farm income. In the process, agritourism provides visitors with entertainment, recreation, participation, and education. Nature tourism consists of commercial operations working toward their visitors' enjoyment, understanding, and appreciation of natural areas while at the same time conserving the local ecological, social, and cultural values and enhancing the local economy.

Called "agriturismo" in Italy and referred to as "sleeping in the straw" in Switzerland, agritourism is well established in Europe, New Zealand, and Australia. In the United States, interest in and desire for information about agritourism and nature tourism is growing. *Agritourism and Nature Tourism in California* was created in response to this demand.

Over 2.4 million visitors experienced agritourism attractions at California's farms and ranches in 2008, according to a recent survey conducted by the UC Small Farm Program (Rilla et al. 2011). They stayed at guest ranches in the Sierra foothills, picked peaches in the Sacramento Valley, played in corn mazes up and down the state, shopped at on-farm stands along the coast, held weddings in fields and vineyards from the coast to the Sierra, and participated in the myriad activities included in agritourism.

Although agritourism and nature tourism are growing industries, agriculture is not. Farmers and ranchers often struggle to survive in today's global marketplace. The pressure of urbanization and shrinking profits has led to California farmers' quest for alternative approaches to maintaining profitable agricultural enterprises. Increasingly, farmers and ranchers are choosing agritourism as a strategy to reap additional income from their existing resources.

Agritourism and nature tourism also provide an opportunity for rural advocacy. With more than 85 percent of Californians living in cities of at least a million people, our state's residents are increasingly removed from the land. Most city dwellers have neither the knowledge nor the personal experience to make informed decisions about product purchases or to vote on policies that impact agriculture. Yet by their sheer number, they profoundly affect rural California.

Agritourism and nature tourism can help this problem. These enterprises can turn tourists into strong allies of family farms and ranches and of the wildlife, habitat, and open space these operations conserve.

Developing agritourism and nature tourism opportunities into money-making enterprises isn't simple. With visitors come issues of liability, public safety, public health, and animal well-being. But farmers and ranchers with vision, skills, and commitment can create an enterprise that responds quickly and directly to tourists' interests and desires.

The authors hope that *Agritourism and Nature Tourism in California* helps nurture these endeavors. It is written to help farmers and ranchers determine their tourism potential and understand the steps of establishing a tourism enterprise. Designed to be a workbook with hands-on activities that can help landowners consider the feasibility of adding a tourism enterprise to their farm or ranch, this book is also a resource for people working with ranchers and farmers.

Agritourism and Nature Tourism in California is divided into seven chapters, arranged from general to specific. The final chapter, "Resources for Success," provides an expanded list of references and Web sites. All chapters provide Web site addresses and reference information. Please note that Web site addresses may change; if an address given in this book is no longer active, a Web search for the name of the site should help you find the new address.

Chapter 1

Consider the Possibilities

Chapter Goals

The goals of this chapter are to help landowners

- ☼ recognize the important role the tourism industry plays in today's global and local economy

- ☼ look at the facts behind agritourism and nature tourism and the potential that agritourism and nature tourism offers to farmers, ranchers, and rural communities

- ☼ learn tactics that can sustain agricultural livelihoods, lifestyles, communities, and the land that supports them all

- ☼ stimulate their interest in operating an agritourism or nature tourism enterprise

- ☼ identify their own operation's prospects for a successful tourism enterprise

- ☼ understand the visitors they will serve and their qualities, interests, and needs

Introducing Agritourism and Nature Tourism

Tourism is the business of providing tours and services to people who travel for pleasure. Rural tourism, then, is the business of providing tours and services to people who travel to rural areas for pleasure. It includes resort stays, agricultural tours, off-site farmers' markets, and other leisure and hospitality businesses. Although rural tourism takes place in the country, it does not necessarily always involve farmers or ranchers.

There are at least two types of rural tourism that do include farmers and ranchers: agricultural tourism ("agritourism") and nature tourism. Taking place on farms and ranches, agritourism and nature tourism diversify agricultural operations and can generate additional income. At the same time, they involve thoughtful care of the land.

Opportunity Knocks

Agritourism and nature tourism are rapidly growing industries. They draw on our increasingly urbanized population, rising incomes, the search for "down-to-earth" fun, and declining recreational opportunities near home. As one agritourism specialist said, "Many urban residents long to experience more of the countryside than they can see from the highways, yet few have friends and family who are farmers. They also want to know more about locally grown land and agricultural practices" (Tosetti 2001). Clearly, many tourists want the very qualities that farms and ranches possess. As a result, opportunity knocks at the doors of farmers and ranchers who have vision, skills, and commitment. This opportunity unites tourists' pursuit for well-being and rural experiences with farmers' and ranchers' needs to supplement their income. It provides operators an opportunity to diversify and protect themselves from fluctuating markets and can also allow landowners to benefit financially from wise use of their land.

Consider Your Tourism Opportunities

For-Fee Recreation Activities

Recreation can provide sustainable and profitable opportunities if the land-owner takes good care of the land.

Access to water bodies, natural areas, and scenic sites

Archery

Bicycle riding and rentals

Bird-watching and other wildlife viewing

Camping

Canoeing and kayaking

Corn or tall-grass mazes

Cross-country skiing

Fishing: catch and release, fly casting

Gathering mushrooms, wild berries, plants, flowers, rocks, gems

Grass sledding

Hang gliding

Hiking

Horseback riding

Hunting: junior hunts

Ice skating

Mountain biking

Nature photography

Off-road biking

Picnicking

Rock climbing and rappelling

Scenic trails: walking, jogging, cross-country skiing, hiking, snowshoeing, horseback riding

Shooting range: firearms, moving-target skeet

Snowmobiling

Snowshoeing

Stargazing

Tobogganing and sledding

Tricycle maze for children

Tricycle racing

Tubing on rivers, ponds, lakes

Turkey shoots

Good for the Land and the People

Indeed, wise land use is as important to farms, ranches, and rural communities as it is to tourism and the general public. Sustainable land practices help build sustainable economies. When rural residents take good care of the land—making decisions that strengthen options for future generations—they help ensure their own long-term well-being. Everyone benefits from sound stewardship.

Agritourism and nature tourism offer farmers and ranchers a creative way to continue operating while caring for the land that supports them. These industries can diversify agricultural operations and provide enough supplemental income to make the difference between profit and loss. They can diversify and help stabilize rural economies, promote economic development, foster community well-being, and build community empowerment. They can also encourage landowners to fully use their land, facilities, equipment, and talents to develop local expertise, provide off-season income, sell products across markets, strengthen existing markets, and establish new markets.

Perhaps the greatest long-term benefit offered by agritourism and nature tourism is the opportunity they provide for rural advocacy. These industries present landowners the chance to teach visitors the values and benefits of farms and ranches.

More than 85 percent of California residents live in cities of at least a million people. Most city dwellers have neither the knowledge nor personal experience to make informed decisions about product purchases or to vote on policies that impact agriculture, but by their sheer number, they can profoundly affect rural California.

In 2008, over 2.4 million visitors participated in educational and recreational experiences at California's farms and ranches, according to a recent survey conducted by the UC Small Farm Program (Rilla et al. 2011). They stayed at guest ranches in the Foothills, picked peaches in the Sacramento Valley, played in corn mazes up and down the state, shopped at on-farm stands along the coast, held weddings in fields and vineyards from the Sierra to the coast, and participated in the myriad activities included in agritourism.

Agritourism and nature tourism can help educate visitors, making them allies of family farms and ranches and of the wildlife, habitat, and open space these operations conserve.

What Is Agritourism?

Agritourism is a specific kind of rural tourism. It is a business conducted by farmers or ranchers on their working agricultural or horticultural operation

can produce livestock and crops, or you can stay clear of traditional production. You can produce new and unique farm goods, or you can add value to traditional products. You can practice alternative farming or ranching (organic farming, for instance), or you can operate via conventional means.

In short, agritourism offers you the chance to operate and share an agricultural operation of choice—whether it's a state-of-the-art, labor-intensive, resource-intensive enterprise or a simple lifestyle experience. Whatever the preferences, whatever the season, prospects for agritourism and nature tourism beckon.

What Is Nature Tourism?

Nature tourism is another kind of rural tourism that consists of a commercial operation working toward its visitors' enjoyment, understanding, and appreciation of natural areas while at the same time conserving the local ecological, social, and cultural values, as well as improving the local economy. It is not ecotourism per se, which targets the nature enthusiast but often disregards local sustainability ("eco"-tours, lodges, and cruises, for example). Rather, nature tourism is a type of ecotourism that focuses on nature-based travel experiences while striving for long-term rural well-being.

As a result, nature tourism is as important to rural residents as it is to tourists. It can help farmers and ranchers care for their land and their communities, calling attention to natural habitats, natural scenic areas, natural resources, and special places.

Many landowners are working to enhance and market natural habitats and wildlife populations. Even where just part of their land is uncultivated, they're trying to attract wildlife and thus visitors. Some have planted native plants. Some have built demonstration areas. Some have even moved beyond the boundaries of their own property and perspectives—for if wildlife knows no boundaries, why should they?

What they're discovering is that collaboration works. Together, many rural landowners, along

for the enjoyment and education of visitors. It is intended to promote farm products and generate additional farm income, in the process providing visitors with entertainment, recreation, participation, and education. Agritourism includes the interpretation of the natural, cultural, historical, and environmental assets of the land and the people working on it.

As an indicator of the growth in this endeavor, when Merriam-Webster, publisher of the Collegiate Dictionary, announced their new words for 2006, "agritourism" was one of only two new business words out of a hundred selected for the year!

Many agritourism operations exist nationwide, including those that provide

- new business opportunities
- supplemental income
- new or unique markets
- educational opportunities
- tourism opportunities
- market-driven land stewardship
- environmentally sound rural development

Agritourism offers great potential. Your imagination—and in some cases government regulations—is all that limits you. For example, you

Demonstrations

Farms and ranches are increasingly being used to demonstrate crafts and traditional rural activities. Farmers and ranchers have many skills to share.

Bread making

Cattle drives, roping, branding

Cooking, canning, wine making, smoking fish and meats

Cow and goat milking

Creating wearable art

Dairy and milking technologies

Flour milling

Furniture making

Gardening

Grain threshing

Jam and jelly making using native or locally grown berries and fruits

Lumberjack skills

Natural horsemanship skills

Organic food production

Planting and harvesting technologies

Quilting

Rug making

Rural crafting: woodworking and willow furniture building

Sheep and cattle herding

Sheepdog and cattle dog demonstrations

Soap making

Solar cooking

Sun-drying food

Sustainable food production

Syrup making

Threshing grain

Weaving

with representatives of local parks, reserves, and wildlife areas, are developing regional wildlife tourism destinations. For instance, where one rancher's duck pond attracts just a few visitors, combining it with other local sites can create quite a stir. Add a neighbor's nesting barn owls, a bat colony under a trestle, and the state Department of Fish and Game's seasonal wetland to the duck pond, and the rural community has something to market!

Weldon, California, offers one example. Weldon lies in rural Kern County, in the Kern River Valley. Traditionally a farming, ranching, and logging community, it has experienced a number of attempts to bolster its declining economy, including wildlife conservation programs and wildlife viewing projects.

The first endeavor was the Kern Valley Bioregions Festival in 1995. That year, history buffs, Native Americans, nature lovers, outdoor-recreation enthusiasts, entrepreneurs, and public agency representatives joined forces to sponsor a festival celebrating the area's diverse cultural, historic, natural, and outdoor recreational heritage.

That festival attracted hundreds of tourists to Weldon and nearby Kernville. Five years later, the 2000 Kern Valley Bioregions Festival brought in more than six hundred nature tourists who spent $61,000 during the three-day period. As a result, the two local chambers of commerce now recognize the economic benefit of wildlife and wildlife conservation. In 2010, they will host their sixteenth annual festival.

Weldon is just one of many nature tourism endeavors under way. Across the rural countryside where farming, ranching, logging, and mining economies have declined, residents are thinking about sustainability. With conservation groups and government

Considering the Opportunities
Case Study: Liberty Hill Farm

More than two decades ago, Beth and Bob Kennett contemplated their farm's financial troubles and chose to diversify by deciding to provide year-round accommodations to tourists in the region. They soon found themselves in the farm vacation business, and no one was more surprised than they.

The Kennetts own and operate Liberty Hill Farm, a working dairy in the heart of Vermont's Green Mountains. Located near Rochester, their farm lies within three hours of Boston and six hours of New York City. This means that more than sixty million people live within six hours of Liberty Hill Farm—many of whom long for a taste of the country.

Naysayers declared that the Kennetts' plans would never succeed: year-round accommodations would never fly because Rochester appealed to mainly winter skiers. But Beth and Bob didn't listen. In 1984, their first guests pulled in the driveway, the first of their annual visits. In 1993, Beth and Bob grossed $25,000. In 2002, they grossed $100,000 and greeted more than twelve hundred guests.

Today, people come to Liberty Hill Farm just to experience a farm and farm family. They marvel at the classic red barn. They delight in the registered Holsteins, numerous farm animals, and big vegetable gardens. Children flock to the tire swing, and everyone enjoys the 150-year-old farmhouse with its eighteen rooms that include a large kitchen, seven guest rooms (and four shared bathrooms), and space to relax.

Of particular pleasure is their hosts' hospitality. The Kennett family embraces its guests. For example, Beth nurtures a daily ritual long lost in many homes: she prepares sit-down, eat-together breakfasts and suppers. And it's not just guests who enjoy this socializing. "We have had dinner guests from every continent and every walk of life," says Beth. "This business gives us a chance to learn and also an opportunity to tell our story about legacy, environmental stewardship, and life on the farm."

Beth and Bob especially encourage families to visit. Running a farm that thrills children and adults alike, the Kennetts urge guests to join in on chores and community activities. From their labor flows fun and education. Writes one visitor in Beth's guest book, "Thank you for opening your home and life to us, and for giving us the opportunity to rethink our values." Another visitor writes, "The best part is our kids will grow up with these wonderful memories of days spent on the farm."

Indeed, many families return year after year—sometimes several times a year and sometimes for a week at a time. New visitors show up based on friends' urging. The Liberty Hill Farm Web site, http://www.libertyhillfarm.com, television interviews, magazine articles, and newspaper stories also draw guests. "My biggest competitor to the Liberty Hill Farm experience is Disneyland. And usually once the family does that, the children want to return here to see real animals!" laughs Beth.

Mary Holz-Clause, Codirector, AgMRC, Iowa State University

agencies, they are beginning to work toward sustainable industries. Their efforts are creating a flow of information and understanding among themselves and also between agricultural operators and the general public. Sensible land practices are helping to generate sensible policies, regulations, and laws. Wise land use is paving the way for long-term community well-being.

Consider the prospects. Explore the activities from which many farms and ranches already are gaining profits.

Travel Trends

After looking at opportunities for farm and ranch diversification, it's time to consider consumer trends and travel trends. When you understand the choices tourists make, you can better plan a successful agritourism and nature tourism enterprise.

Tourism Pays!

The numbers of travelers, dollars spent, revenue generated, and jobs created by tourism increase annually statewide, nationwide, and worldwide. In 2010, tourism and travel composed nearly 9 percent of the global economy (ILO 2010). The number of "person-trips" in the United States increased as well, though numbers are down during the 2009 economic downturn. (One person-trip is a trip of more than fifty miles from home for reasons other than work or school.) In 2008, Americans took 1.987 billion person-trips, down 0.6 percent from 2007 (OTTI 2010).

As the recessions of 2008 and 2009 indicated, this is still a good time for local and unique travel trips closer to home.

And States Profit

Due to its large population and its diverse attractions, California has the largest tourism industry in the United States, with 338 million domestic visitors traveling to and through California in 2008, 86 percent of these being other Californians. That same year, tourism contributed $97.6 billion to California's economy, providing 924,000 jobs and generating $5.8 billion in state and local taxes (CTTC 2008).

California agritourism operators surveyed by the UC Small Farm Program estimated that 88 percent of their visitors in 2008 were from California, with 50 percent coming from the same county as the farm or ranch they were visiting (Rilla et al. 2011). This finding is consistent with state tourism and travel commission figures of

CORN MAZE ENTRANCE

Exhibits and Lectures

Exhibits and lectures about the land and nature mesh well with farm and ranch settings. Imaginative landowners are offering facilities for numerous topics.

Animal husbandry

Aquaculture

Conservation technology

Environmentally sensitive farm and ranch technologies

Farm and ranch antiques, tools, and equipment

Forest ecology

Herbal medicine

Historical crafts

Irrigating, fertilizing, and integrated pest management

Local community: history, culture, modern living

Local natural history

Native plants

Rural crafts: quilting, doll making, fruit basket building, and toy building

Soil conservation technologies

Sustainable agriculture

Wildlife management

---

86 percent for the same year. Only about 3 percent of visitors were from Canada or other foreign countries.

Growth in Agritourism

Agritourism is growing nationwide as farm operators in many states offer agritourism activities as one way to diversify and increase their on-farm profits (see Brown and Reeder 2007).

Other national data sources also support the economic development potential of agritourism. Nearly two-thirds of all U.S. adults (87 million) took a trip to a rural destination from 2002 to 2005 (Geisler 2011), and more than 82 million people visited farms during a one-year period in 2000 and 2001, including approximately 20 million youth and children under the age of sixteen (U.S. Forest Service 2003). Additionally, the economic impact of the wine industry has been surveyed, with $2 billion attributed to tourism-related sales in California (Wine Institute 2006).

Growth in Nature Tourism

Did you know that nature tourism is one of the fastest-growing segments of the travel market? According to the International Ecotourism Society, nature tourism has averaged a 20 to 30 percent increase each year beginning in the early 1990s (TIES 2006). In 2006, the National Wildlife Watching Survey conducted by the U.S. Fish and Wildlife Service (USFWS) reported that 87.5 million Americans over the age of sixteen spent $29.2 billion to observe, feed, and photograph wildlife: "These activities composed the second most popular pastime in the nation (after gardening)" (Leonard 2008). California has the highest number of wildlife watchers in the country, with 6.2 million wildlife and nature tourists spending more than $4.6 billion on these endeavors annually (Leonard 2008).

What all of this means to farmers, ranchers, and rural community members is economic potential. Nature tourists seek the amenities that farmers and ranchers can provide. Each year, nature tourists spend more time on the road and more time watching wildlife. In their pursuit, they buy gear, food, transportation, and lodging. And they spend cash on guide fees, public land use fees, magazines, membership dues, contributions, land leasing, land purchases, and habitat improvement, according to the 2006 USFWS report (Leonard 2008).

Bird-Watchers Mean Big Business

Within the nature tourism industry lies the fastest-growing sector of the nature tourism business—bird-watching. It is more popular than hiking, camping, fishing, or hunting. It's even more popular than golf, reported Fortune magazine, with Americans preferring bird-watching to golfing as a way to unwind during vacation (USFWS 2000). According to the National Birding Survey conducted in 2006, the average birder is fifty years old and more than likely has a better-than-average income and education (Leonard 2008). The higher the income and education level, the more likely a person is to be a birder. This survey presented information on the participation and expenditure patterns of 48 million birders in 2006. Trip-related and equipment-related expenditures associated with birding generated over $82 billion in total industry output, 671,000 jobs, and $11 billion in local, state, and federal tax revenue. This impact was distributed across local, state, and national economies.

In short, bird-watchers represent a significant market segment that farmers and ranchers can attract to supplement their income. When

---

Although the economic impact of agritourism has not been thoroughly researched, a variety of recent state surveys have indicated its importance to the local farm economy. In Vermont, a 2003 survey revealed that one-third (2,200) of farms received an average of $8,900 from agritourism activities in 2002 (Beus 2008). In California, half the operators responding to a state survey reported less than $10,000 in revenues for 2008, while 21 percent (55) had agritourism revenues of $100,000 or more; one-third of the operators had annual gross profits of between $10,000 and $99,000 (Rilla et al. 2011).

The annual Space Coast Birding and Wildlife Festival attracted over 4,500 participants in 2010 and contributed almost $1 million to the local economy in Florida (Space Coast Living 2011).

The Eagles and Agriculture event (tours and workshops) in Carson Valley, Nevada, is a collaborative effort initiated in 2004 to demonstrate how agricultural stewardship benefits wildlife and wildlife habitat in Nevada. The Carson Valley agricultural and ranching community, along with the Carson Valley Visitors Authority, the Audubon Society, the Great Basin Birding Observatory, along with the conservation community, cosponsor the unique combination of viewing birds of prey and agricultural education.

Lured by photography and birding workshops, participant surveys reveal that people are more interested in ranching, history, and agriculture than with bird-watching by a margin of three to one. Attendance has doubled from 628 the first year, and annual revenues are close to $20,000 (Garrison et al. 2005). Over three thousand people have attended previous years' events and have given rave reviews. All proceeds benefit local wildlife, conservation, and agriculture projects.

Participant Experiences

In general, agritourists and nature tourists want to participate in activities. They can choose from an endless array of hands-on activities.

- Animal birthing
- Animal shows
- Bread baking
- Barn raising
- Cheese-making workshops
- Ethnic dancing
- Farm or ranch vacations: branding, roundups, chuck wagons, haying, farming, cattle drives
- Farm school for children
- Food tasting
- Garden plot rentals
- Grow your own: fruit, vegetables, flowers, crops, Christmas trees, fish
- Hay rides
- Herbal medicine making
- Make your own old-fashioned toys, clothes, Christmas decorations, food
- Mushroom log seeding
- Natural habitat development: bats, birds, fish, snakes, insects, game
- Pick, harvest, and cut your own fruit, vegetables, flowers, crops, Christmas trees, fish, firewood
- Pond construction
- Pony rides
- Quilt making
- Pumpkin patch and corn maze
- Rent a tree, bush, garden, flowerbed, crop
- River fording
- Rug making
- Trail rides, including overnight trips
- "Wild West" trail rides with bandits

farmers and ranchers cultivate healthy habitats, they bring in birds that bring in bird-watchers who bring in dollars.

In 2010, there were thirty-seven birding and wildlife festivals listed under "festivals" at the California watchable wildlife Web site, http://www.cawatchablewildlife.org.

Counties Benefit

Tourism plays a significant role not only in California's state economy but also in its counties' economies. Each county has scenic, cultural, and historic attractions of potential interest to visitors. Local tax receipts totaled $2.2 billion in 2008, ranging from $15.5 million in rural Mono County to $29 billion in Los Angeles (CTTC 2008). In addition, the travel and tourism industries make up the fourth-largest employer in California, supporting nearly 924,000 jobs and continuing to add them while thousands of jobs have been lost in the manufacturing, information, and trade sectors of the state (CTTC 2008).

Consumer Trends

Growth in a state's tourism industry translates to growth in its agritourism and nature tourism industries. For consumer trends, log onto the Tourism Division of California's Technology, Trade, and Commerce Agency Web site, http://www.visitcalifornia.com. There you'll see that visitors' pursuits and interests fall in line with farmers' and ranchers' amenities. As you consider the opportunities ahead of you, keep tourist trends in mind.

Consumer Trends in America

Tourists, not surprisingly, like to shop, and then they like to participate in outdoor activities, visit historical places and museums, visit national and state parks, and take part in cultural events and festivals. Furthermore, they like to participate in activities, not just to watch them. Increasingly, Americans want to experience something new, connect with their roots, and touch the land. As a result, tourists are ever more involved in educational experiences like nature tourism, heritage tourism, ethnic festivals, and adventure travel. Americans have more to do and less time to do it.

Long, extended vacations have been replaced by short, intensely active vacations, making key markets a two- to three-hour drive away. For ranches and farms located near other outdoor recreation amenities and businesses, locale is a plus. According to the 2007 Tourism Industry of America (TTIA) Ideal American Vacation Report, the ideal vacation destinations for American vacation travelers are those that offer an easy travel experience, a sense of fun and adventure, and local flavor.

Tours

Tours generate income not only from entrance fees but also from food, crafts, and souvenirs. They can be scheduled weekly, monthly, or by appointment. To attract more tourists—for a package price—they can be scheduled in conjunction with other operations' tours and other points of interest.

Aquaculture operations

Bird and wildlife sanctuaries

Cider mills

Conservation activities

Croplands

Culture

Ethnic sites

Farm and ranch buildings

Fish ponds and farms

Flower and herb farms

Food processing plants

Forest sites

Historic sites or buildings

Horse logging ventures

Hydroponics operations

Land restoration sites

Local and rural communities

Maple syrup production facilities

Natural areas

Natural habitat

Orchards

Sawmills

Scenic attractions

Specialty livestock operations: angora goats, llamas, dairy

Traditional farms and ranches

Wetlands

Wineries

Consumer Trends in California and Beyond

Most people who travel in California are on vacation. For example, three-quarters of all California person-trips in 2008 were leisure—trips that entailed short, frequent journeys within the state and, often, within the traveler's local area. The visitor and convention bureaus in Sonoma, Marin, Placer, Sierra, and Plumas Counties direct their advertising efforts to Bay Area and Sacramento travelers in particular and California travelers in general.

Geotourism, a traveler preference for destinations that protect the authenticity and geographic character of place, is growing, according to The National Geographic Society's Center for Sustainable Destinations, http://travel.nationalgeographic.com/travel/sustainable/. Geotourists want authenticity. Visit the National Geographic Web site for virtual map guides of unique areas of the country. New virtual map guides have been released for the North Coast (http://www.visitredwoodcoast.com) and the Sierra Nevada (http://www.sierranevadageotourism.org).

Visitors are seeking alternatives to standard motel and hotel lodgings. Other trends include shopping, touring the local countryside, hiking, bicycling, and visiting small towns, festivals, and fairs. Agriculture-related food festivals, fresh cuisine, and the long-held California tradition of wine tasting also are much appreciated.

Generational Travelers

The agritourism and nature tourism industries serve primarily senior citizens and middle-aged "baby boomers." Baby boomers are entering their sixties and will soon compose 30 percent of the national population. Travel tops their list of desired retirement activities. They are looking for self-fulfillment and exploring their self-potential—and, accordingly, they provide an ideal target market for tourism products that offer education and challenge. Baby boomers are "doers," not sitters. The aging population is not just a U.S. phenomenon: by 2020, there will be an estimated seven hundred million people over age sixty worldwide.

Senior citizens currently compose 20 percent of the national population, and those over fifty-five represent 80 percent of vacation dollars spent in the United States (TTIA 2004). Their interests revolve around independence and self-sufficiency, social and spiritual connectedness, selflessness, personal growth, and revitalization. They seek serenity, comfort, cleanliness, and value.

Typically white and well-educated, senior tourists are interested in museums, historical sites, and cultural exhibits and activities. Many participate in passive activities like walking, visiting historical sites, viewing wildlife, observing nature, and bird-watching; fewer participate in physically demanding outdoor recreation. Of special note is that intergenerational travel—grandparent with grandchild—is a growing trend, and the experience provides memories and a legacy for both individuals.

Farm, Ranch, and Community Entertainment

Rural entertainment provides fun, education, and fond memories.

Bonfires

Clam bakes

Dancing

Education

Guided nature tours

Hands-on experiences: face painting, painting, pottery, woodworking workshops, plays

Hay, sleigh, and tractor rides

Historic interpretation

Interactive games

Music concerts

Mystery theater

Participatory plays

Playhouses

Plays

Pumpkin carving

Socials for dining

Storytelling

Talent shows

Most tourists in California are Californians, comprising 86 percent of in-state travel (CTTC 2008). In addition, California is the top travel destination in the United States, receiving 21 percent of all overseas visitors. Therefore, one of the biggest potential tourism markets are the various urban ethnic groups in California. It is important to cater to them.

Cultural and Heritage Tourists

Family-friendly activities and cultural interests rank in first place as travel decision influencers among cultural or heritage travelers. Cultural tourism involves museums, art galleries, concerts, and plays. Heritage tourism embraces the culture and history of an ethnic group—whether it's Native American, European American, Asian American, African American, or any other ethnic background. Together, these industries draw millions of visitors to California.

Nearly half of all adult Americans planning a vacation intend to go to a historical site. American and foreign tourists alike visit historical sites; local, state, and national parks (which often have cultural and heritage sites); festivals and crafts fairs; and museums, plays, and concerts. These visitors want to learn.

The number of foreign visitors is rising, too, and they are also interested in pursuing new and "real" experiences. At the same time, it is interesting to note the attraction that "going western" has for foreign travelers residents. For example, Germany has an estimated four hundred "western" clubs with forty thousand total members (Lopinto 2009). These visitors spend time and money studying the American West and Native American history, riding horses, learning to shoot a bow and arrow, and cooking over a campfire. "As children we used to play cowboys and Native Americans," the *New York Times* reported a German tourist visiting Oregon as saying (Egan 1998). As tourists, these visitors come to California for the "real thing":

outstanding scenery, good value, and cultural and educational opportunities.

Families

Families vacationing in California generally seek a number of short trips rather than one long vacation. Family travel is for leisure, and family leisure falls into four main categories: a visit with family and friends, a special event, a weekend getaway, and a vacation. Although many families take trips during their children's summer breaks, a growing number of families travel during the school year.

Like other agritourists and nature tourists, families seek participatory experiences, preferring new outdoor activities at new places. The average number of children in a family is 1.7, and the average household income is $58,925 (ERS 2009). In search of bargains, families look for affordability, safety, proximity, and lodging. They want to experience a healthy environment and the rural lifestyle, and they want their children to learn where food and fiber come from.

Childless Travelers

Childless travelers have more disposable income and take longer trips than do families with children. Accordingly, they spend more money on vacations. Today, 66 percent of American households contain no person younger than eighteen, and more than one-quarter of American households contain only one individual (U.S. Census Bureau 2010). One of the best opportunities for tourism marketing is "vacations for one."

Affinity Travelers

These people want to travel together as a group. They include groups of women knitters, church groups, volunteer groups, and senior citizen groups. These travelers aren't simply looking for workshops or another museum tour. They want to link up with others who share something deeper: the same religion, race, gender,

or lifestyle. The types of trips available are as varied as the affinity groups who take them. One agritourism group is organizing food and wine and farmstead cheese tours. Your farm could be a destination.

Nature Tourists: Of Special Interest

Nature tourists have their own profile worth considering. They are environmentally conscious people who enjoy wilderness settings, wildlife viewing, hiking, and trekking. Typically thirty-five to fifty-four years of age, they consist of men and women in roughly equal numbers. They seldom travel with children and take vacations of eight to fourteen days. They spend an average of twelve days per year watching wildlife.

Nature tourists tend to be upper-middle-class Americans: urban, affluent, highly educated, sophisticated, busy, and middle-aged. Those who travel outside of their home state are generally of higher income than vacationers within their home state and the national population itself. That is, nearly half of nonresident nature tourists have an annual income of at least $50,000; nearly one-third have at least four years of college education, and 65 percent live in cities with a minimum population of 250,000 (Wood 2002).

Like agritourists, nature tourists are looking for a quality experience. They appreciate educational experiences and activities that deepen their viewing skills and their understanding of the local area. They like on-site interpretative exhibits and brochures, plant and animal identification courses, and wildlife study.

Nature tourists also seek hands-on experience. Rather than passively observing plants and animals, they will spend money and effort to help build and restore sites. They often want active adventure as well as intimate encounters with nature.

Other Outdoor Recreation Enthusiasts

Agritourism and nature tourism enterprises draw people with all sorts of outdoor interests. An operation can contain one or

more of the many amenities that outdoor enthusiasts appreciate, as indicated by the following information published by the Outdoor Industry Foundation (2007).

Hunters and anglers

Over 7.4 million California residents and nonresidents sixteen years old and older fished, hunted, or watched wildlife in California in 2006. Of the total number of participants, 1.7 million fished, 281,000 hunted, and 6.3 million participated in wildlife watching, spending a total of $3.4 billion on trip-related expenses. Hunters and anglers tend to be men from sixteen to thirty-four who live in rural areas. They love the outdoors and love being outside. They value solitude and appreciate unpopulated areas. Worried about today's increasingly limited access to fish, animals, and land, many anglers and hunters are looking for a reservation system or a place that provides land access (Wood 2002).

Adventure enthusiasts

Adventure tourists want thrills, excitement, and challenge. They climb mountains, bicycle the backcountry, shoot rapids, and participate in other sports that challenge their physical or mental condition. Often very social people, adventure tourists usually travel in large groups. They tend to be young and healthy, with higher-than-average incomes. Today, they select their adventures from over ten thousand adventure-trip operations and spend more than $110 billion each year. More than half of the U.S. travelers have taken adventure trips.

Mountain bicyclists

The number of off-road bicyclists—known as mountain bikers—is climbing 20 percent each year. Today, the market contains approximately 10 million people. Most are single or childless men from twenty-one to thirty-two years old. As more trails on public land are closed to bicycles, the demand escalates for trails on private land.

Facilities for People

Farms and ranches often possess buildings, scenery, and surroundings appropriate for many public activities.

Assisted-living accommodations

Business meetings

Business picnics

Church groups

Country weddings

Family reunions

Fish ponds

Heritage enactments

Historic train rides

Interpretive programs

Playgrounds

Retreats

Reunions

Trails: foot, horse, vehicle, off-road vehicles

Wildlife viewing platforms

Youth camps

Hospitality Services

Because of the serenity they often provide, farms and ranches are sought for lodging and other hospitality services. They can supply numerous services.

Assisted-living services

Bed and breakfasts

Cabin rentals

Catered functions: weddings and reunions

Childcare

Farm and ranch vacations: cattle round-ups, chuck wagon meals, cattle drives

Horse livery stables

Interpretive centers

Meeting and banquet facilities for business and pleasure: country weddings, retreats, reunions, picnics, parties

Pet boarding and training

Pet daycares

Picnics and trail lunches

Restaurants serving dishes made from products from farms, ranches; local and wild products

School field trips

Storage for boats and vehicles

Youth camps

Wagon trains

Campers

Campers are primarily young people from twenty-five to thirty-four who camp more than once a year. Families also enjoy camping because it provides time for parents and children to experience the outdoors together.

Hikers

Hiking is rapidly growing in popularity. Participation is related to income and age, and it is significantly more popular among high-income households and younger Americans. Participation is not related to gender, however: equal numbers of men and women hike. Typically keen environmentalists, hikers like to enjoy nature and improve their physical fitness all at the same time.

Winter enthusiasts

Millions of Americans enjoy snow activities. High-income earners between twenty-four and thirty-six years of age dominate these sports, which include snowboarding, skiing, sledding, ice skating, cross-country skiing, and snowshoeing. In particular, people of households earning more than $100,000 per year are three times more likely than people of lower-income households to participate in winter sports (Wood 2002).

Horseback riders

Horse enthusiasts usually belong to higher-income brackets. They want riding lessons, trail riding, and arena riding, and they also enjoy cattle drives.

Gardeners

More than half of adult Americans garden as a hobby. Gardeners are often in pursuit of knowledge. They enjoy public gardens, private gardens, and gardening education. Such activities can be enjoyed simultaneously with sightseeing, walking, and studying and photographing nature, as well as relaxing, learning, and establishing a sense of connectedness with the earth.

Culinary enthusiasts

The tremendous growth in farmers' markets in California and beyond stresses consumers' desire to purchase and eat locally and seasonally. Consumers want to plan and prepare their meals with fresh, high-quality produce. This growing trend is a good sign, since many of our most profitable agritourism operations in California sell on-farm fruits, vegetable, and value-added products.

Michael Pollan's 2006 bestseller *The Omnivore's Dilemma* also shows how the slow food and local food movement has been growing for years (see the Slow Food Web site, http://www.slowfood.com). Many families and individuals are interested in growing or knowing where their food comes from—they aren't

Alternative Crops and Value-Added Products

High-value nontraditional crops, specialty livestock, and farm-related services are providing farmers and ranchers with supplemental income.

Access to water bodies, natural areas, and scenic sites for recreation, education, and research

Alternative-livestock products: goats (meat, milk, cheese, soap); llamas, alpaca, and angora goats (wool or breeding stock); free-range livestock; pastured poultry and livestock; rabbits (meat, fur)

Aquaculture: fish, clams, shrimp

Bait: minnows, worm farming

Bison, elk, deer

Boarding, training, caring for horses, hunting dogs, cats, and other pets

Building wood products from locally sawed lumber and by-products

Canned, smoked, dried, and other preserved goods

Christmas trees

Craft sales: dried flowers, wreaths, furniture

Cricket farming

Farmers' markets

Firewood cutting and sales

Fish farming

Flowers and herbs

Fruit and nut orchards

Game dressing: ducks, geese, pheasants

Gift shops: arts and crafts made on farm, on ranch, in local area

Grapes and by-products: jellies, wreaths, wood, and wine

Ground-cover production

Guide service: hunting, fishing, pack trips, sightseeing, photography

Hay sales

Horse boarding

Horseback riding

Jellies and jams from native wild berries

Lavender: fresh, decorated, dried, crushed

Mushrooms: shiitake and others

Nursery products: indigenous species, native plants, shrubs, annuals, nursery stock

Organic poultry and livestock

Organic vegetables

Potpourri

Roadside market farm stand: products produced on farm, ranch, or locally such as fruits, vegetables, nuts, cider, Christmas trees, greenhouse plants, herbs and flowers, meats, eggs, firewood, dressed game

Pick or cut your own (U-pick): fruits, vegetables, flowers, Christmas trees

Sawdust and wood shavings sales

Sesame production for oil or condiment

Sod farming

Straw sales

---

simply interested in purchasing the food, they want to meet the farmer who grew it, visit the farm or ranch, and perhaps even develop a personal relationship. These experiences can provide an agritourism experience that people are willing to pay for. Culinary enthusiasts are also interested in food, fine dining, and new eating experiences. They appreciate wine, handcrafted beers, and gourmet cooking, preferring something new to something they've had before. Culinary enthusiasts tend to be from forty to fifty or older and often travel without children. They appreciate educational opportunities. They want to meet local people, try local specialties, and watch particular foods being made. As a result, cooking class vacations and gourmet tours are another important trend for agritourism marketers. The nationwide organization Chefs Collaborative works with chefs and the greater food community to celebrate local foods and foster a more sustainable food supply. See their Web site, http://chefscollaborative.org, for a local chef connection.

The Food Network and other food-themed programming have helped boost interest in food and culinary travel. Michael Coon, director of the Culinary Institute of America's Worlds of Flavor travel programs, believes that television programs and an increased interest in educational travel have a lot to do with it (Hunter 2006). He adds, "I would say people are looking for an alternative to lying on the beach and sort of having more educational elements to their vacations versus relaxation."

Why Vacation?

Agritourists and nature tourists vacation away from home, in the rural countryside. The Agri-Business Council of Oregon's Agri-Tourism Workbook (2007) reported the following reasons people vacation away from home, ranked according to importance.

To build and strengthen relationships

The primary reason Americans travel on vacation is to spend time alone with their family. They want to be together with their family in stress-free surroundings, and they consider a trip away from home to be the ideal opportunity. They view travel as a time to rekindle and strengthen their relationships. Many Americans also view travel as a time to start new friendships, and they look for social interaction throughout their trips.

To improve health and well-being

Vacations are vital to travelers' physical and mental well-being, both for individuals and families. Furthermore, research shows that most California visitors want to participate actively in outdoor activities.

To rest and relax

Americans on vacation want to rest and relax. A trip away from home is a trip away from work and worry. When they return, they feel refreshed and renewed.

To experience adventure

Some travelers vacation away from home to find adventure. They want their vacations to provide excitement, be it dangerous or romantic.

Alternative Marketing

There are many innovative ways to market crops. Numerous techniques can increase farm and ranch sales.

Community-supported agriculture

Direct marketing

Direct selling to schools and restaurants

School field trips

Farmers' markets

Internet sales to distant buyers

Pick or cut your own: fruits, vegetables, flowers, Christmas trees

Rent-a-tree, berry bush, garden, or flowerbed

Roadside sales

To escape

Most tourism surveys indicate that many people travel to escape their daily routine, worry, and stress, and to attain what they sense is missing in their lives (Krippendorf 1986). They seek something different: perhaps a better climate, a slower pace of life, cleaner air, prettier scenery, or quieter surroundings.

To learn

Better-educated travelers reported that they travel to learn and discover. They want to see, hear, touch, and feel unfamiliar things. More specifically, they want to learn or practice a language, study a culture, explore gourmet foods or wines, or investigate spirituality.

To mark a special occasion

Many Americans vacation away from home to celebrate life milestones and special occasions. New relationships, marriages, birthdays, and professional achievements provide a reason. These people usually travel with loved ones, creating memories that last a lifetime.

To save money or time by traveling locally

Tourists sometimes take short, local vacations to save money or time. Indeed, both money and time limit nearly every vacation decision. Some vacationers are very frugal.

To reminisce

Another reason Americans travel is to relive fond memories. Some vacationers—particularly older ones—visit a farm to rekindle memories of the simple rural lifestyle they once knew. Although these people buy food, lodging, transportation, and souvenirs, they in fact are purchasing a sentimental journey.

To view nature

In addition to the above survey information, a recent U.S. Fish and Wildlife survey reported that, on average, nature tourists view wildlife to observe nature's beauty, relax from daily pressures, get away from home, and be with friends and family (Leonard 2008). These tourists like learning about nature, being physically active, and meeting people with similar interests. Social interaction and relaxation is particularly important, sometimes secondary to seeing wildlife.

Points to Remember

- The tourism industry is a primary force in the world economy, and it is an increasingly important industry in the United States in general and California in particular.

- Agritourism and nature tourism provide farmers and ranchers a chance to diversify their operations, supplement their incomes, improve their communities, and care wisely for the land.

- The agritourism and nature tourism industries are consumer-focused enterprises that respond quickly and directly to consumer needs, preferences, interests, and values.

- Agritourists and nature tourists have particular needs, interests, preferences, and values that must be understood and addressed if alternative farm and ranch enterprises are to succeed.

- Agritourism and nature tourism enterprises are diverse, achievable, and potentially profitable, limited only by imagination and—in some cases—government regulations.

Acknowledgments

The sections "Travel Trends" and "Consumer Trends" were adapted in part from B. Garrison et al., Development and Marketing Strategies for Birding and Wildlife Tourism in the Greater Reno, Nevada Region (Nature Tourism Planning Web site, 2005). The "Consider Your Tourism Opportunities" sidebars were adapted from Alternative Enterprises for Healthier Profits, Healthier Land (USDA Natural Resources Conservation Services Information Sheet AE-1). The information in the section "Why Vacation" is adapted from the Agri-Tourism Workbook (Portland: Agri-Business Council of Oregon, 2003, http://www.aglink.org/contact.php).

Chapter 2

Evaluating Your Resources: Is Tourism for You?

Chapter Goals

The goals of this chapter are to help landowners

- identify how a tourism enterprise can mesh with their operation's current philosophy and goals
- consider the costs and benefits of an agritourism or nature tourism enterprise
- recognize the features their farm or ranch offers visitors
- work toward concrete agritourism and nature tourism opportunities

Three-fourths of all agritourism operators are motivated by their desire to increase farm profitability, according to the 2009 University of California Small Farm Program survey of California operators (Rilla et al. 2011). Over half of operators making more than $50,000 during 2008 found their venues to be profitable, with pumpkin patches and on-farm sales of products as the most profitable types of activities for them.

Revisit Your Goals

Planning, persistence, and patience are key to a successful tourism enterprise. Therefore, before you embark on your new endeavor, take time to plan. Early planning will get you thinking about the costs and benefits of a tourism enterprise.

First, examine your business goals. Identify your business's underlying philosophy and objectives. Analyze your operation's current capabilities and situation. Consider how a tourism venture can contribute to your and your family's future.

Next, sit down with your business partners and family members and discuss your business and personal goals. Look at where you are now and where you want to be. It is essential that you obtain full agreement from everyone involved in your operation's decision making, both family members and outside interests. Should you skip this step, you're liable to face such problems as misunderstandings, disagreement, conflict, and personal and interpersonal stress, and experience only limited success in your new tourism operation.

For a more detailed discussion of the planning process, see chapter 3, "Creating Your Business Plan."

Consider the Costs and Benefits

Agricultural and nature tourism can supplement your farm or ranch income—but there is no guarantee. Small businesses have high failure rates. Therefore, you must understand the tourism business and its impact on your lifestyle and time. To succeed, you must enjoy people and interact well with them, maintain high standards, market effectively, and make sound financial decisions.

Equally important, you must take time to review the costs and benefits of the agritourism and nature tourism industries (see table 2.1). Remember that you are entering the business of agritourism and nature tourism. This entails personnel management, record keeping, marketing, retailing, accommodations, entertainment, food service, catering, and interpretation. Soon, you'll find yourself amid such time-consuming tasks as learning about regulations, hunting down financing, purchasing insurance, advertising your business, paying for its upkeep, looking for visitor hazards, supervising employees, and doing paperwork, all the while trying to meet your family's needs. Clearly it's wise to know what lies ahead before taking on an agritourism project.

Identify Your Competition and Customers

With your family and other business partners, check out your local competition. Find out who it is, where it is located, what attractions and activities it offers, and who it hosts. Visit as many agritourism and nature tourism operations as possible. Talk to successful operators about their businesses and experiences and your plans. Book a room or a tour, and maybe volunteer or work there. Observe, and don't forget to ask questions!

You need to consider your customers as well as your competition. Chapter 6, "Designing Your Marketing Strategy," will help you identify the visitors you want to attract, but take time to think about them in advance. Ask yourself who would be attracted to your farm or ranch as it is today. Who might you attract with your new tourism venture in place?

The number of potential customers for agritourism and nature tourism operations is high. However, using a "shotgun" approach to draw them is expensive and rarely effective, and a "build it and they will come" approach is economically precarious. Instead, choose and target a customer sector within the agritourism and nature tourism industry. There are people who are curious about how your farm operates and interested in what your farm offers. Identify these people as your potential clientele, with their own specific interests, preferences, needs, and values.

Table 2.1 Benefits and costs of an agritourism and nature tourism enterprise

Benefits	Costs
Provides potential additional income.	Provides a low financial return, at least at first.
Creates a physical operation that appreciates in value.	Interferes with farming or ranching operations.
Efficiently uses underutilized facilities, equipment, land, and talents.	Hard work! Adds workload to family members.
Allows you to be your own boss.	Demands your full and constant attention, interfering with family time and activities.
Allows you to work your own hours.	
Allows you to express yourself creatively.	
Allows you to live your own creation.	Steals your privacy—people are always around.
Is personally rewarding.	
Generates new opportunities for spouse and children.	Requires you always to be "on"—upbeat, available, and attentive.
Maintains family attention and interest on the farm or ranch.	
Provides the opportunity to meet people—visitors as well as agritourism and nature tourism professionals.	Involves risk and liability.
Provides the chance to play a significant role in community activities.	Can create staffing problems.
Provides the chance to educate people about rural living, nature, and the agriculture industry, which in turn can lead to improved local policies.	Generates excessive paperwork.
Provides the chance to learn about outside perspectives, which in turn can lead to better-educated rural residents and improved local policies.	
Promotes the agriculture industry.	
Models sustainable local industries.	

Who Visits?

Agritourism operators estimated that on average 88 percent of their visitors in 2008 were from California, with an average of 50 percent from the same county as the ranch or farm they were visiting (Rilla et al. 2011).

To begin to identify your visitors, create a profile of your ideal customers. You can start by completing the form in figure 2.1.

Figure 2.1

Who Are My Ideal Customers?

Where do my ideal customers come from? _____
_____

What are their ages? _____

Are they single or married? _____

Are they families with children? _____

What is their ideal party size and composition?
_____

What is their income? _____

What magazines, newspapers, and books do they read?
_____

What are their hobbies and interests?
_____

How much time do they have for the activities that I offer?
_____

What characteristics do they have?
_____

- ☐ Are they looking for demonstration and guidance?
- ☐ Are they seeking relaxation?
- ☐ Do they want entertainment activities?
- ☐ Do they want action or physical activities?
- ☐ Other

Assess What You Have

Now take a long, hard look at your farm or ranch. It is financially prudent to launch your new enterprise with what your operation possesses today rather than what you go out and purchase. So consider the amenities you own. What is intrinsic and unique to your land and operation? Why would a person enjoy visiting it? What memorable experiences would a customer take with them?

Think about the natural environment, scenic areas, and history of your operation and its surroundings. Consider the aesthetics of your property and the logistics to reach it. Determine the distance from main roads, proximity to urban areas, and proximity to active tourism regions. Look at access, infrastructure, services, and facilities.

Think about your "people skills." Good people skills are fundamental to success in the tourism business. You should enjoy working with people, whether visitors or employees. In addition, you and your employees should have a sense of humor, a dedication to excellence, a strong work ethic, and physical stamina.

The worksheets in this chapter can help you identify your operation's amenities and thus evaluate your options and ideas. Chapter 6, "Designing Your Marketing Strategy," provides more detail.

Review Your Physical Resources

Location plays an important role in the success of your enterprise. In the worksheet in figure 2.2, consider your operation's physical resources.

Consider Your Operational and Management Skills

Success calls for good operational and management skills. Use figure 2.4 to consider your perspectives, assets, and weaknesses.

Assess Your Personal Skills

What are you best at? Fill out the worksheet in figure 2.5 to assess your own skills.

Rate Your Human Resources

What about the people visitors might have contact with—your staff, neighbors, and local community businesses? Will they strengthen your business or impede its success? Make a list of your human resources in figure 2.6.

Take a Look at Wildlife

To the nature tourist, habitat and wildlife matter. What plants and animals do you have on your farm or ranch that people would want to see? If you are a private landowner and want to restore fish and wildlife habitat on your property, the Partners for Fish and Wildlife Program may be for you (see http://www.fws.gov/cno/partners/). The California Native Plant Society Web site, http://www.cnps.org/, allows you to locate your specific geographic area and identify common plant types found there. Your local Natural Resource Conservation Service, Resource Conservation District, or Cooperative Extension office may also be an excellent resource.

Map Activity Sites

What are your ideas for visitor activities? Contemplate where on your property you'll hold these activities and how you'll address your visitors' needs. Sketch out the locations of your activities in the box provided. Take into account the amenities you've identified throughout this chapter. Figure 2.7 is an example to get you started, and in the box below is room for you to plot your own activities. Be creative!

Figure 2.2

What Are My Physical Resources?

LAND

Legal Description

How much land do I own or have access to? List the acreage amount, location, and proximity. List
property that is deeded, leased, private, state owned, and federally owned. _____
_____
_____

Land Use

Access to a roadside may enhance your ability to sell produce directly from the farm. Farms or ranches
with wooded areas can be used for mushroom production or hunting. Open areas might be good sites
for demonstrations, classes, weddings, or retreats. Fallow fields might provide income from goose
hunting, and pastures might be used to graze alternative livestock. List your land's current use including hayfields, rangelands, croplands, and feed grounds. _____
_____
_____

Land Features

What does my land look like? Land that is unique or beautiful can provide income opportunities from
farm tours, hiking, horseback riding, nature tours, and photography and art tours. List your land's
features including its elevation; topography; access to public roads; and natural places like woodlands,
meadows, wetlands, and water bodies. _____
_____
_____

Soil Type

What are the characteristics of my soil? What is it best suited for? _____
_____
_____

Water Bodies

What streams, lakes, rivers, and ponds will attract tourists? Land with water can be used for such activities
as fishing, bird-watching, photography, duck hunting, and water sports such as canoeing or kayaking.
_____

CLIMATE

How will weather patterns affect the activities I provide? What is the growing season? Will I need to
irrigate? _____
_____
Temperature: monthly average and variation_____
Rainfall: monthly average and variation _____
Snow: ground cover periods and accumulation depth _____

FARMSTEAD FEATURES

What buildings, fences, corrals, working facilities, equipment, roads, and paths are on my land? Is my
home well-suited for visitors or is the barn a possibility for conferences or workshops?_____
_____

HISTORICAL RESOURCES

What special historical or cultural buildings and features exist on my property or nearby? _____
_____
_____

ADDITIONAL ATTRACTIONS

What other resources do I have on my land that might attract tourists? Consider livestock, fishing
areas, vistas, and proximity to natural or created points of interest, for example. _____
_____

LOCAL INFRASTRUCTURE

What local infrastructure exists? Consider roads, local transportation systems, traffic, parking, signage,
nearby lodging, and dining, for instance. _____
_____

Resource Evaluation Case Study

Hunewill Ranch

It's a story all too familiar to ranchers and farmers today, one involving financial crisis and personal stress. Near the eastern border of Yosemite National Park, Napoleon Bonaparte Hunewill founded Hunewill Ranch in 1861. The cattle ranch encompassed 4,500 acres of meadows and forests near Bridgeport, California, and 1,250 acres of winter range in neighboring Smith Valley, Nevada. It included a 160-acre gold claim as well—plus grazing rights to additional lands. With all of these resources, by the 1930s, not a dollar could be earned.

The Great Depression had reduced cattle prices to five cents per pound and, accordingly, decimated the operating capital at Hunewill Ranch. There was no cash and little hope. Descendants Stanley H. and Lenore Hunewill had only creativity, determination, and their spectacular ranch setting when they built three cabins and advertised pack trips, trout fishing, and hunting opportunities. When clients from the coast began to filter in, their guest ranch was born.

Today, Hunewill Ranch is still owned and operated by the Hunewill family. With its breathtaking mountain backdrop, short distance to urban California, and outstanding customer service, the ranch has earned a stellar reputation among guest ranches. But guest ranching is still just part of the operation. There is also a 1,200-head cattle ranch and a grass-leasing business as well, which together support four families.

It's a winning combination, the families say. While the guest ranch allows them to hold onto their land and raise cattle, the cattle ranch provides guests an education and authentic atmosphere. Grass leasing fits right into this plan. It helps stabilize the ranch's income, and—to the conservation-minded Hunewills—it ensures ecological integrity.

Because Hunewill grass-leasing fees are based on weight gain, the families need high-quality forage to ensure high weight gain and high grass-leasing fees. Consequently, they rotate their pastures, which has stabilized their stream banks, improved forage production, and increased trout populations. In turn, their healthy countryside draws nature lovers again and again.

Indeed, approximately 70 percent of Hunewill Ranch guests are return guests. Many of these return people refer new customers. Each week of the summer, forty-five to fifty-five people come to the ranch to commune with nature, soak up the scenery, photograph wildflowers, fly-fish in the heart of the eastern Sierra, and, most of all, ride horses.

Horses are Hunewill Ranch's specialty. Guests can ride one of the ranch's 120 horses or bring their own. Whether novices or experts, they're guaranteed a variety of experiences. They can lope through meadows, trail ride into the mountains, or amble through autumn colors. They can gather and sort cattle, bring in herds for vaccinating or branding, doctor cattle in the field, or move cattle between pastures and between summer and winter ranges. Or, they can take horseback riding lessons and play horseback riding games. At Hunewill Ranch, there's something for everyone, right down to sing-alongs and square dancing.

The Hunewills realize that ranch guests have helped them sustain their lifestyle and land. "Guests are a gift," says Betsy Hunewill, acknowledging that even now, seventy years later, guests give the family as much as the family gives to them. The ranch's Web site can be found at http://www.hunewillranch.com/.

Evaluate Your Financial Resources

You will need financial resources for business start-up and upkeep. While some alternative enterprises have high start-up costs, others require little up-front investment. Do you have access to loans or other sources of capital? Use the worksheet in figure 2.3 to consider your operation's financial resources.

Figure 2.3

What Are My Financial Resources?

_____

What are my start-up costs?

What access do I have to capital?

_____

_____

_____

_____

_____

Consider Your Operational and Management Skills
Success calls for good operational and management skills. In the form below, consider your perspectives, assets, and weaknesses.

Figure 2.4

What Are My Operational and Management Assets?

LANDOWNER'S AND MANAGER'S STRENGTHS AND GOALS

What intangible assets do I have to help create a farm or ranch recreation operation? Consider interpersonal skills; marketing ability; knowledge about specific topics such as livestock management, gardening, local history; and special skills such as horseback riding, furniture making, and cooking.

_____

_____

_____

FAMILY STRENGTHS AND RESOURCES

What "intangibles" can my family members bring to this enterprise?

_____

_____

_____

NEIGHBOR AND COMMUNITY RESOURCES

What talents, skills, and interests might local residents add?

_____

_____

_____

FARM AND RANCH PERSONALITY

What is the personality of my farm or ranch? Is it serene, vibrant, laid-back, or interactive? This description can help you to choose suitable events and to design them appropriately.

_____

_____

_____

FARM AND RANCH ACTIVITIES

What current farm or ranch activities might appeal to the public? Remember that what you consider routine might be unusual and interesting to the nonfarming public. Be creative! These activities might include cattle drives; calving or lambing; trail rides; roadside produce stands; machinery operations such as planting, cultivating, and harvesting; and on-site food processing.

_____

_____

_____

Assess Your Personal Skills

What are you best at? Fill out the following worksheet to assess your own skills.

Figure 2.5

What Are My Personal Skills?	No	Somewhat	Yes
Do I like meeting and working with all types of people?			
Do I like to entertain and serve strangers?			
Am I patient and sensitive to the needs of visitors?			
Do I have the physical stamina and vigor to maintain my operation while properly serving customers?			
Do I know the natural history of my area?			
Am I			
a self-starter?			
willing to take responsibility?			
organized?			
able to make and carry out decisions?			
able to solve problems?			
Do I have experience with budgeting?			
planning?			
managing people?			
communicating?			
presenting my ideas to many people?			
selling?			
keeping financial records?			

Think "Quality"

Agritourists and nature tourists are looking for a first-rate experience. Superb quality is the most important element you can offer, declares one agritourism operator: "Everything must be of the same high quality. If anything is at a lower level, the entire experience will be brought down. People remember the little things they didn't like" (Friemuth 2001).

What is a high-quality experience? It doesn't mean a four-star hotel. It means that visitors participate in meaningful, fun, authentic activities and attractions. It means they are welcomed, respected, and cared for. And it means they get what they came for, and, when they leave, they want to return.

Make Access Easy

A high-quality experience begins with clear directions to your farm or ranch. After you decide on visitor areas, clearly mark visitor access, using entrance and simple directional signs. Keep driveways cleared and graded, and set up parking near the activity site. If guests will arrive in the dark, light your signs and driveways. A remote location may be much valued—but it must be easy to find. Read chapter 5, "Developing Your Risk Management Plan," for more ideas.

Look Good

A high-quality experience helps your visitors feel comfortable on your property. Keep your grounds clean and your gardens well groomed. Paint fences and plant flowers. When possible, renovate buildings rather than bringing in modular structures. Where farm odors permeate, consider posting an educational sign that explains the issue.

Rate Your Human Resources

What about the people visitors might have contact with—your staff, neighbors, and local community businesses? Will they strengthen your business or impede its success? Make a list of your human resources below.

Figure 2.6

How Supportive Are Other People?
STAFF: FAMILY AND HIRED
Do they have special talents or abilities?
Do they have time to deal with the public?
Are they willing to deal with the public?
Are they friendly with the public?
MANAGEMENT TEAM
Are they committed to working with me to make a successful venture?
NEIGHBORS
Are there any conflicts?
Will they allow bordering access?
Will they support recreation options?
SHERIFF AND DEPUTIES
Do they support my proposal?
Concerns or barriers
GAME WARDENS AND BIOLOGISTS
Do they support my proposal?
Concerns or barriers
GOVERNMENT OFFICIALS AND REGULATORS (LOCAL, STATE, FEDERAL)
Do they support my proposal?
Concerns or barriers
ECONOMIC DEVELOPMENT STAFF (LOCAL, COUNTY, REGIONAL, STATE)
Do they support my proposal?
Concerns or barriers
BUSINESS AND TOURISM ASSOCIATIONS
Do they support my proposal?
Concerns or barriers
OTHER LOCAL BUSINESSES
Do they understand and support my proposal?
Concerns or barriers
Would they be interested in collaborating with me?
How can I reduce my hurdles to success?
Additional thoughts

Offer Hands-On Activities

A high-quality experience allows your visitors to be directly involved in activities if they so desire. Create an environment that invites participation. For example, provide the opportunity for visitors to hand-feed calves or do self-directed activities such as fishing and canoeing. Activities needn't be full-length vacations; one-hour tours and half-day excursions are a great beginning.

The choices you make depend on your goals, philosophy, and resources. As you consider options for your guests, capitalize on natural settings, scenic areas, and history. For instance, if you own an organic farm identify it in your signage and promote it in your advertising. If your farm or ranch borders an old mining claim, the El Camino Real, a stagecoach trail, or a Maidu Indian path, tell that story. Be accurate with your information! Do your homework before you open your doors and monitor your family and employees to ensure quality presentation skills and accuracy. See chapter 6, "Designing Your Marketing Strategy," for more ideas.

Contemplate where on your property you'll hold these activities and how you'll address your visitors' needs. Sketch out your activities' locations. Take into account the amenities you've identified throughout this chapter. Figure 2.7 is an example to get you started, and in the box below is room for you to plot your own activities. Be creative!

Go the Extra Mile

Provide a valuable experience or quality product.

Always do a little more than the customer expects.

Provide solitude! Because solitude in many state and national parks is now very difficult to find, it is highly marketable.

Provide natural settings. The more natural amenities a site possesses, the less development and monetary risk exists.

Ensure security and safety.

Make reservations available online. Some public-sector recreation opportunities are so limited that a waiting list of several months is common.

Serve special-interest groups, like senior citizens, physically and mentally impaired citizens, urban residents, photographers, or students.

Offer "friends of the farm" memberships and special events or other benefits to regular or returning customers and on your Web site.

Obtain feedback. Ask your guests about things you want to know; their perspectives are key to success. Ask them what they like about your enterprise, what they don't like, and what they wish you'd offer. To successfully gain their views, consider the methods used by some businesses:

- In clear view, place a large suggestion box with a sign asking, "How can we better serve you?" or "What is it you need, want, or miss?"

- Train employees to ask visitors for comments and suggestions and to write them down.

- Act on your visitors' suggestions. Respond to their ideas by keeping a blog or customer feedback on your Web site so they know you're listening.

Figure 2.7

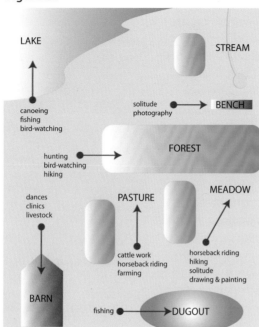

Points to Remember

- Lots of thought and planning precede the successful agricultural and nature tourism enterprise.

- Before pursuing any agritourism or nature tourism enterprise, you must make sure the potential enterprise will mesh with the philosophy and goals of your farm or ranch.

- Agritourism and nature tourism have advantages and disadvantages, both of which require serious consideration.

- Before pursuing a tourism enterprise, identify and understand your competition, market, and existing farm and ranch resources.

Make your own sketch here.

Chapter 3

Creating Your Business Plan

Chapter Goals

The goals of this chapter are to help landowners

- understand the business planning process

- complete each component of a business plan

- recognize how to integrate agritourism and nature tourism enterprises into their existing operation

- link the business plan to other information this book offers

- pursue financing for their new enterprise

- cultivate a flourishing agritourism or nature tourism venture

What Is a Business Plan?

Your business plan clarifies your values, goals, challenges, and operations. When drafted properly, it helps you obtain financing, if needed, for growth or property improvements, and it provides guidance and focus for managing your operations. In short, a business plan allows you to make mistakes on paper and gives you the chance to work through business decisions before committing resources.

A business plan gives partners a clear understanding of the agritourism or nature tourism enterprise, facilitates business management, and assists in getting financing. One agritourism operator from the state agritourism survey called his business plan the "nuts and bolts" of initiating his enterprise. "It includes data to help me decide if this idea is really feasible and lenders decide if they should advance money to this enterprise," he explained.

The results of a California agritourism survey completed in 2009 show that those with business plans for their entire farm were about twice as likely as those with no business plans to have agritourism incomes more than $100,000 (Rilla et al. 2011). Keep in mind that your business plan is fundamental to obtaining financing—so be realistic with projected figures, justify the need, and be enthusiastic but reasonable about your tourism enterprise.

What Is Included in a Business Plan?

An important point to remember is that a business plan is a dynamic, ongoing process. Like a tractor, it's a tool. Pieces wear out and are replaced; implements are added as needs change; an overhaul is occasionally required. And, like a tractor, the business plan has many parts. The following paragraphs describe a business plan's components in a suggested—but not required—order.

Executive summary

The executive summary is a one-page summary of your plans. The first part consists of your business idea, and the second consists of the conclusions you've provided in your financial strategy. The executive summary should come first in your business plan, but it should be written last.

Mission statement

The mission statement reflects the core purpose of your business, stating your values and goals in a focused sentence less than fifty words long.

Your business concept or idea

Your business idea comprises a one-page, concise, complete, and persuasive statement that describes the what, where, why, and how of your enterprise. It includes a description of your

- agritourism or nature tourism enterprise
- products and/or services
- target audience
- enterprise's fit with others in the region, both competitive and complementary

Measurable goals and objectives

Goals are the long-term plans you want to achieve in your agritourism enterprise in the next three to five years, both professional and personal. Objectives are your short-term (one year or less) plans that identify how you will meet those goals. Goals are what is to be accomplished; objectives are how goals are to be accomplished.

Background information

The background information about your enterprise should be drawn from industry research and market analysis. It explains why you are selling your product or service. For example, it might contain estimates of the number of potential visitor days and expenditures, visitor preferences and interests, your competition, and the complementary services in your area. See chapter 1, "Consider the Possibilities"; chapter 6, "Designing Your Marketing Strategy"; and chapter 7, "Resources for Success," for additional information.

Management needs and management history

The management component of the business plan explains how you will run your business. It describes your management team and their experience, as well as your legal structure, insurance, and staffing needs. It also notes the regulatory agencies you will be working with and how you will address their requirements.

Marketing strategy

Your marketing and advertising strategy details how you plan to promote your operation. Your strategy might be simply relying on word-of-mouth promotions or repeat customers, or it might be a well-planned promotional campaign with marketing methods, distribution, and location. Marketing might include Web sites and social networking like Facebook or Twitter, as well as electronic messages, printed materials, mass mailings, radio spots, newspaper advertisements, and other media advertising. See chapter 6, "Designing Your Marketing Strategy," for more information.

Financial strategy

Your financial strategy identifies your sources of existing debt and your financing needs, specifying your fixed assets, start-up costs, and several basic forecasts, as well as your monthly principal and interest payments. It also includes a conclusion that explains how your new enterprise will fit into your current operation and summarizes your financial documents. These financial documents are your profit and loss statement with assumptions; balance sheet, including assets, liabilities, and net worth; and cash flow projection, including sales projection and assumptions.

Appendix

The appendix, located at the end of the business plan, furnishes supporting documents. It includes your financial statements, customer support statements, and credit terms available to your business.

Establish a Mission Statement

Discover Your Values

One big difference between your agritourism enterprise and other agritourism enterprises is the values that you and your family hold dear and how those values are expressed in your enterprise. Your values compose the very core of who

3G Family Orchard

Background Information

- This operation is a family-owned orchard, inhabited and run by three generations of family members.

- The orchard is mixed pomes and stone fruits.

- All family members value keeping the land in the family, but only one child of the third generation is interested in farming it.

- Family members remain very involved in their community and church.

- Family members value their privacy and their right to use the cultural practices they view necessary to producing high-value, marketable crops.

- The third generation believes the family must diversify its operation to ensure economic viability amid rising competition in the traditional marketplace.

Selected New Enterprises

- Direct-market sales of early and heirloom varieties.

- On-site in-season restaurant and pie shop.

- Value-added processed products.

Mission Statement

3G Family Orchard will preserve the heritage of our farm and fruit-growing region by selling the highest-quality fresh fruits at traditional farm stands, farmers' markets, and pie shops. We will educate people about the importance of family farms by participating in the Farm Bureau and community planning processes.

you are. Through them you filter your business decisions. If you make decisions that ignore your values, you face consequences that conflict with your fundamental being.

Clearly it is important that you identify your values. Because farms and ranches are home-based and traditionally involve family, all members of your family should examine their values. Everyone's participation is critical when you contemplate a new business direction—particularly one that involves visitors to your home.

Chapter 2, "Evaluating Your Resources," advised that you sit down with your family to learn everyone's view about the proposed enterprise. Now meet with them again to consider everyone's

values, goals, and resources. Then you can decide which tourism opportunities best fit your family and your operation.

Photocopy the worksheet in figure 3.1. Each family member should fill it out and add important items. Then combine everybody's answers on one form.

Now ask yourselves:

- What do we want this property to look like at the end of our stewardship?

- What quality of life is important to us?

- What relationship do we want with family members? The local community? Our customers? Vendors?

- What changes are we willing to make in our farm or ranch?

- What changes are we willing to make in ourselves?

- Who can help us make the above happen? (Refer to your answers from the "Assess What You Have" section in chapter 2.)

- What resources do we have to make the above happen? (Again, refer to your answers from chapter 2.)

Write Your Mission Statement

From the values you've established and the enterprise you've selected, you can now write a mission statement. Your mission statement should reflect what you want to do, how you want to do it, and how you will evaluate what you've done. Among other things, it requires that you ask yourself, What is the purpose of our proposed enterprise?

What are its benefits? Who are its customers? A mission statement requires serious financial deliberation as well, with you carefully considering how you can earn enough money to keep doing the fun!

For a better understanding, look at these three fictitious examples.

- 3G Family Orchard: a three-generation family orchard operation
- Working Landscapes Ranch: a third-generation cattle ranch left by grandfather to grandson
- Grandpa's Farm: an educational operation run by career corporate managers who retired early and bought the land outright to raise truck-farm crops, citrus fruits, and a small herd of goats

It's your turn now! With your family, complete the worksheet in figure 3.2.

Figure 3.1

Our Values

Values	Least Important	Somewhat Important	More Important	Most Important
Spending time with my family				
Embracing spirituality or church				
Enjoying my peace and privacy				
Participating in our community				
Building a stronger local community				
Keeping the land in our family				
Maintaining our heritage/homestead				
Restoring our farm or ranch				
Protecting the resource base				
Enjoying our natural environment				
Displaying our land stewardship				
Making land available for our kids and grandkids				
Leaving a legacy				
Continuing the family business				
Ensuring our economic sustainability				
Producing food for others				
Providing the highest-quality product				
Providing the most affordable product				
Growing unique crops or animals				
Teaching others				
Continuing my education				

Working Landscapes Ranch

Background Information

- This operation is a third-generation cattle ranch on which only the third generation (grandson) lives.
- The grandson inherited the land.
- The grandson was raised off-ranch but each spring helped his grandfather move cattle to the higher summer range.
- The grandson values the land's natural beauty.
- The grandson requires economic sustainability to pay taxes and keep up the ranch.
- The grandson wants to restore the historic ranch buildings.

Selected New Enterprises

- Nature tourism and education.
- Guided fishing in the trout streams that cross the property.
- Bird-watching.
- Cross-country skiing.

Mission Statement

Working Landscapes Ranch will provide sound ecological stewardship by offering guided long-weekend and week-long ranch stays to select groups of active adults who appreciate natural beauty and low-impact outdoor recreation in a rustic setting.

Goals and Objectives

Mission Statement

Working Landscapes Ranch will provide sound ecological stewardship by offering guided long-weekend and week-long ranch stays to select groups of active adults who appreciate natural beauty and low-impact outdoor recreation in a rustic setting.

Goals

1. Native plant grasses and forbs proliferate on our property.
2. Native shrubs and trees stabilize stream banks, provide shade, and lower water temperatures for native steelhead.
3. Hillside erosion is limited.
4. Viable populations of native wildlife species indicative of a healthy ecosystem exist on the land and in the water.

Objectives

- Use rotational grazing to increase the native grasses and forbs.
- This coming spring and fall, reintroduce native shrub and tree species to stream banks.
- By next summer, develop a stepped ponding system in the north end to reduce sedimentation of the streambed and pasture flooding in years with heavy rain.
- By next spring, build opportunity stream crossings for cattle to mitigate bank damage.

Action Steps

- From March through May of this year, increase the land's water-holding capacity by fencing off the stream beds and using solar panels to pump water to pasture troughs in the north valley, using volunteer labor of family and local high school students.
- Contact the Natural Resources Conservation Service and the Department of Fish and Game to learn about this region's ecosystem and our "ideal" habitat and to determine indicator species (plants and animals) and their needs.
- Contact local resource agencies to partner in reestablishing natural, historic creek channels and develop a management plan.
- Apply for conservation programs that may mitigate the costs of this endeavor, including stream and pond reestablishment.
- Work with the California Rangeland Trust to sell a conservation easement, capitalize pasture and stream restoration, and provide short-term living expenses.
- Lease pasture to Barker Ranch in exchange for help with cross-fencing.

Source: Adapted from Rosenzweig 2001.

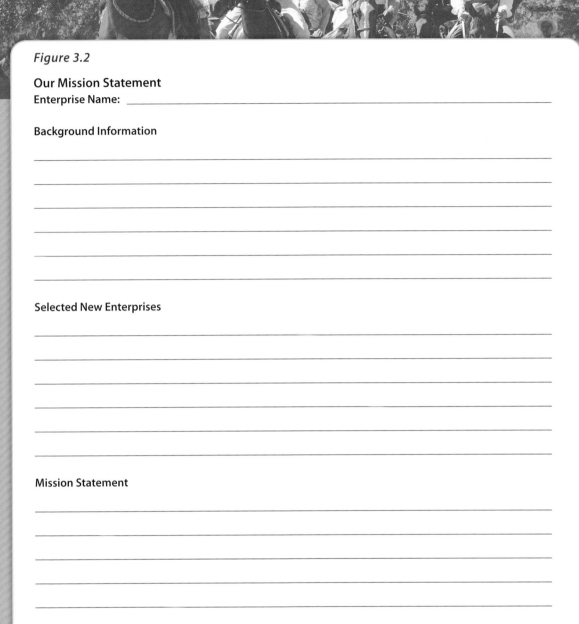

Figure 3.2

Our Mission Statement
Enterprise Name: _____

Background Information

_____
_____
_____
_____
_____
_____

Selected New Enterprises

_____
_____
_____
_____
_____
_____
_____

Mission Statement

_____
_____
_____
_____
_____
_____
_____

Describe Your Idea

After generating your mission statement, describe your business idea. Your business idea—or "business concept"—is the what, where, why, and how of your enterprise. It is a particularly flexible component of your business plan, changing throughout the plan's creation whenever unexpected assumptions or costly plans arise.

Write this business concept as if you were explaining your tourism venture to your brother-in-law, who is a lawyer, at a family reunion. You want to be succinct, complete, and—because you might need a loan—persuasive.

Your business concept is a one-page summary, written in the first person ("I" or "we"). It serves as the first part of your executive summary. In it, you describe the enterprise, product or service, clientele, and how your enterprise fits with others in the region. Take a closer look at the following four topics.

Your agritourism or nature tourism enterprise

Explain when your enterprise was started or is projected to start. Describe why it was started, who started it, and how it fits into your overall farm or ranch operation. Identify your business or legal structure.

Your product or service

Write down what you are selling to the public. Explain why people should buy your product or service, what it does, how it is unique, what it will cost to provide, and what you will charge. How does your business compare with other similar enterprises?

Your clientele

Who are your targeted customers? Define their gender, age, income, occupation, location, family status, education, and interests (see chapter 2, "Evaluating Your Resources"). How much time will they spend on your operation—an hour, an afternoon, or a week?

Grandpa's Farm

Background Information

- Operated by career corporate managers who retired early and purchased the land outright to raise truck-farm crops, citrus fruit, and a few goats.
- The managers are looking for a sense of community not found in the city.
- The managers are interested in farmers' markets and unusual varieties of vegetables and herbs.
- The managers hope to provide a place where their grandchildren can experience the same things they did on their grandparents' farms.

Selected New Enterprises

- School visits and family day trips.
- Farmers' market participation.
- Seasonal weekend festivals.

Mission Statement

Grandpa's Farm will provide children a safe place to learn where food comes from. We will provide family-centered, seasonal, on-farm activities, invite community schools to visit the farm, and transport our produce and animals to town to reach people who cannot come to us.

Business Concept

Grandpa's Farm will be a mixed specialty vegetable truck farm in Purple Valley, approximately 30 miles from Blue City. We will certify all 40 acres of productive land with the state's Certified Organic Program. We will operate the farm as a sole proprietorship, with help from apprentices from the local community college's agriculture program.

We will grow specialty vegetables on a year-round basis: tomatoes in the summer, specialty mixed baby lettuce in the spring and fall, and specialty potatoes in the winter. We will choose all crops based on their salability as specialty crops to the organic market.

We will prioritize our sales accordingly: 1) to independent retailers (grocery stores with a produce person who has authority to buy) in our county; 2) to local restaurants in our county; and 3) to food-service jobbers in our county (businesses who get orders from food services, buy the product directly from the farmers and terminal brokers, and deliver the order). However, during our first 2 years of production—as the farm builds up—we will sell weekly at the farmers' markets in Red Town, Grayville, and Blue City.

At any given time, 20 of our 40 acres will be in active vegetable production. In the other 20 acres, 5 acres will be in soil building, 5 acres will remain in specialty citrus, 5 acres will be available to a small mix of farm animals for grazing and to establish a "farm atmosphere," and 5 acres will be a learning laboratory for children and families. We will invite school groups to come during the week, and we will hold long-weekend family festivals to emphasize and celebrate the change of the seasons. We will capitalize our farm's start-up with profits from the sale of our city house and from cash from annuities. After 3 years, the farm will be self-supporting and will provide us with a net salary of $40K.

How your enterprise fits with others in the region

What complementary and competitive services exist in your area? Develop and evaluate your products and services with your resources in mind, and then decide whether you'll compete with or complement already existing businesses. Describe your collaborative efforts.

A Sample Business Concept

To better understand how to describe your business idea, look at the business concept in the sidebar "Grandpa's Farm." Remember that Grandpa's Farm is an educational farming operation that raises truck-farm crops, citrus fruits, and a small herd of goats.

Write Your Business Concept

Now it's your turn. With your family, describe your business concept. You can use the worksheet in figure 3.3.

Set Measurable Goals and Objectives

Having generated your mission statement and described your business concept, you now can set measurable goals, objectives, action steps, and tasks. If this sounds daunting, think of the Wizard of Oz.

- Goals are the Emerald City, your destination.
- Objectives are the Yellow Brick Road, the path you must follow to reach your destination.
- Action steps are the large stretches of the road, the specific activities you must undertake to attain your objectives.

- Tasks are each brick in the road, the details.

Once you've drawn the map of your Yellow Brick Road, you can put dates next to sections to see how fast you must travel. In other words, first ask yourself the basic questions: Who? What? When? Why? How? Then set timetables to achieve goals and objectives.

Ask yourself who will be responsible for the various activities on your farm or ranch. Who are the key employees and what are their titles? Will you need new employees or can you retrain existing ones? How long do you plan to operate this venture? How many hours are you willing to commit to it? Where do you see the business in five years? How are you going to promote customer satisfaction? Consider setting goals and objectives for your sales volume, profits, customer satisfaction, owner compensation, and employee training.

To understand more clearly, take a closer look below at goals, expected outcomes, objectives, and action steps. Afterward, read about how Working Landscapes Ranch will set and meet its goals.

A Closer Look

Goals are your desired future. They are visionary, yet they must be attainable and fit well with your mission statement. They respond to the challenges, opportunities, and potentials of the future, yet they vary little from year to year. To write a goal statement, begin with action verbs like "provide," "create," "ensure," or "develop." For example, you might write, "To develop a well-recognized crane-viewing site within the next 3 to 5 years."

Business Plan Case Study
Bonanza Creek Ranch

Just north of Lennep in southcentral Montana lies Bonanza Creek Ranch. Nestled between the Crazy and Castle Mountains, the ranch covers more than thirty thousand acres of sweeping valleys blanketed by grasslands. Since 1877, Bonanza Creek Ranch has been a cattle operation owned by the Voldseths. Now, it is also a profitable guest ranch.

In the mid 1990s, David and June Voldseth decided to diversify and opted to open a guest ranch. They determined that guest ranching could "help keep the ranch solvent and help maintain the wide-open spaces," says June. Moreover, it would provide her the welcome opportunity to meet and mingle with people.

From the beginning, the couple knew their strengths and desires. They wanted to entertain no more than sixteen guests at one time—and so provide an intimate atmosphere and superior customer service. They wanted to offer horseback riding, own quality horses, and hire knowledgeable wranglers. They also wanted to offer a variety of riding experiences, including day-long trips, cattle herding, and horsemanship lessons.

In addition, the Voldseths decided to offer hiking and mountain bike riding on the ranch and in the Lewis and Clark National Forest. They decided to promote trout fishing in nearby streams and two large ponds. They also knew that the Charles M. Bair Family Museum revealed colorful local history. Whatever their guests' leisure-time pursuits, when it came to dining, David and June agreed that only good home cooking would suffice.

That was 1994. To set their plans in motion, David and June drew up a business plan that allowed time for both a learning curve and a deficit. Once the business plan was completed, the Voldseths shopped for lenders, a contractor, and building materials. "It was surprising to see the price differences, especially among contractors," June recalls. When they found the best terms, they took out a fifteen-year business loan, putting up some ranch land as collateral.

Then came construction. Because the family wanted lives separate from the guest ranch, they developed an area solely for guests. At the tree line on a hillside facing the Crazy Mountains, they built attractive kit cabins and a main lodge, all purchased from Lindal Cedar Homes. Each cabin was placed within walking distance of the lodge and corral, and out of sight from all others.

Marketing followed construction, and a whole new learning experience unfolded. Concerned about attracting enough visitors, the Voldseths pursued several marketing techniques, growing increasingly aware that each technique bore its own costs and benefits. David and June created a brochure describing their ranch to send to interested people; they found it ineffective alone but an invaluable supplement. They hired booking agencies including Off the Beaten Path, http://www.offthebeatentrack.com, and GORP Travel, which they felt was a temporary marketing strategy until return guests and referrals could carry business. They joined the Dude Ranchers' Association, http://www.duderanch.org, which guarantees standards and provides publicity and a support system for guest ranchers. In addition, they advertised in Sunset Magazine (ineffective) and were promoted in Gene Kilgore's Ranch Vacations and featured in April 1998's Travel and Leisure (very effective).

Furthermore, the Voldseths developed the Bonanza Creek Ranch Web site, http://wwwbonanzacreekcountry.com, which drew more visitors to the ranch than any other marketing technique. Although it required up-front costs, it now requires minimal upkeep and therefore little expense.

Four years after evaluating their resources and writing their business plan, David and June saw their first profit. Today, their venture earns a stable income and increases the overall cash flow of Bonanza Creek Ranch. It also breaks the family's isolation and broadens their horizons. They found their niche in being small and having good riding horses. They no longer take beginners, as they found that good riders liked to be with other good riders. In 2009, almost 50 percent of their guests were international. The Voldseths learn as much about their guests' lives as guests learn about theirs and have found guest ranching a pleasure and an eye opener.

Figure 3.3
Our Business Concept

Enterprise Name: _____

Business Concept

_____
_____
_____
_____
_____
_____
_____
_____
_____

Then there are expected outcomes. Expected outcomes specify the conditions that will exist when you meet each goal, describing desired conditions, quality levels, and subjective conditions like feelings. They help clarify your goals and help guide and direct the development of annual objectives. They are measurable. You should aim for at least three expected outcomes under each goal, beginning with "This goal will be satisfied when," as in the following example. Next come objectives. Objectives identify how a goal will be accomplished. They are specific, measurable, realistic, and obtainable within 1 year.

Goal 1: To develop a world-class crane-viewing site within 5 years.

Expected outcomes: This goal will be satisfied when we

1. Have a visitors' center providing a panoramic viewing area plus educational materials.

2. Acquire half of our annual clients via word-of-mouth.

3. Receive attendance and recognition from conservation and ornithological groups.

Selected objectives to accomplish this goal:

1. Design our visitors' center with the help of our local architect; to be completed at the end of this 1-year period.

2. Determine our marketing strategy; to be completed during the first 3 months of this 1-year period.

3. Bring members of the local National Audubon Society to our site for advice; outside consultation to be completed during the first 6 months of this 1-year period.

4. Create a Web site; to be accomplished during month 6 through month 12 of this 1-year period.

Action steps are specific activities that must be done to accomplish each objective. They specify what work needs to be done, who will do the work—family or perhaps the National Audubon Society—and when the work will be completed.

An Example of Setting Goals and Objectives

Consider the Working Landscapes Ranch, the third-generation cattle ranch described earlier in which the operation was left by grandfather to grandson. Grandson has established two goals:

Goal 1: To renovate historic buildings. Working Landscapes Ranch must renovate its historic shacks and bunkhouses as the first step in its tourism enterprise. The renovated buildings can provide housing for cross-country skiers in the winter and for ranch hands in the summer. Eventually, they'll provide housing for nature tourists.

Goal 2: To improve the health of the ranch's summer range. Working Landscapes Ranch must restore the health of its land before it can offer nature tourism activities like bird-watching and fishing. We need to better manage summer pastures. Birds will be among the first to colonize the improved habitat, while fish populations will profit from improved stream conditions. Bird-watching and fishing will be among our first nature tourism activities. Land restoration will increase our income from cattle. Better land management means a higher carrying capacity for cattle, with our goal being 300 cow-calf pairs. We can either lease the summer pasture or graze our own herd on it.

Write Your Own Goals and Objectives

Give it a try. With members of your family, establish three to five measurable goals and their expected outcomes, objectives, and action steps. Use the worksheet in figure 3.4.

How Will You Run Your Business?

This section discusses many topics related to the way you run your enterprise: assembling a management team, determining the legal and organizational structure of your business, buying insurance, and coping with all of the management challenges ahead.

What kind of management challenges might you face? There could be issues with employees, regulatory agencies, safety and risk management, and your business's future if you are injured or die.

Write Your Own Goals and Objectives

Give it a try. With members of your family, establish three to five measurable goals and their expected outcomes, objectives, and action steps.

Figure 3.4
Our Goals and Objectives
Enterprise Name: _____

Mission Statement

_____
_____
_____
_____

Goal #__ (To be accomplished in_____)

_____
_____
_____
_____

Expected Outcomes

The goal will be satisfied when

1. _____
_____
_____

2. _____
_____
_____

3. _____
_____
_____

Objectives (How will I achieve this goal)

_____
_____
_____

Detailed Action Steps (Who, What, When)

_____
_____
_____

Identify Your Management Team

Your business management team runs the day-to-day business of your enterprise and makes sure you meet your goals and objectives. It includes paid advisors—people with whom you already have a business relationship or already pay for services, and perhaps individuals within your operation. It could include a lawyer familiar with business law, an accountant, and an insurance agent, along with experts in research and development, marketing, strategic planning, and information technology. A well-rounded management team will furnish you a strong technological, administrative, and institutional knowledge base and will provide alternatives that help you avoid pitfalls. Make sure to list the names and qualifications of everyone on your management team in the appendix of your business plan.

Identify Your Legal Structure

The legal structure of a business has implications for management control, taxation, liability, and estate planning. Deciding the form of ownership that best suits your business venture should be given careful consideration. Use your management team and key advisors to assist you in the process. You will select among sole proprietorship, general partnership, limited partnership, corporation, limited liability company, and limited liability partnership, and, in the process, you will develop an understanding of liability, control, and taxation. See the Small Business Planner at the Small Business Administration Web site, http://www.sba.gov/smallbusinessplanner/, for more information on your options.

Remember that liability is a very complex and broad topic. Even if a business is organized under something other than the sole proprietorship described below, many lenders will require personal guarantees and additional security such as a second deed on the ranch.

You should consider reviewing your decision regarding your choice of legal structure with professional advisors, including your attorney and tax accountant. The decisions are not irrevocable, and they may be staged to occur at different levels of growth. However, to maximize your goals, professional consultation prior to making your decision would be the most effective way to do business.

- **Liability.** Liability is the degree to which your personal assets are exposed to business risks. You can protect your personal assets from these risks by choosing to conduct your business in a form that limits liability.

- **Control.** The form of business you select also determines the degree of control you

have over your business. A corporation with stockholders has a much different control issue than a partnership or a sole proprietorship.

- **Taxation.** A corporation is a separate taxable entity. If the corporation makes a profit, the corporation pays a tax on the profit. It then distributes the profit in the form of dividends to shareholders, and a second tax is paid.

- **Tax benefit or burden.** Three factors determine the overall tax burden of an entity:
   o the tax brackets of the owners of the entity
   o the amount distributed currently rather than retained by the entity
   o the length of time before the owners dispose of their interests in the entity

One of the biggest tax advantages that owners of a regular corporation have is that they may also be employees of the corporation. Thus, stockholders who are employed by a regular corporation can qualify for tax-favored fringe benefits that are not available to the unincorporated business owners.

Sole proprietorship

The sole proprietorship—a business person—is the simplest form of business. A husband and wife are considered to be a single person for these purposes, but the business cannot be owned by more than one person or married couple. A sole proprietorship is viewed as an extension of the individual who owns, manages, and is directly liable for the business. The entity requires no legal documentation; however, a sole proprietor must comply with general business requirements, such as business licenses, Department of Labor regulations regarding unemployment and industrial insurance, and Department of Licensing regulations requiring registration for use of a trade name that does not include the full legal name of the owner of the business. A sole proprietor may apply for an employee ID number. (For a sole proprietor, your social security number is your tax ID number.) For more information, see chapter 4, "Understanding and Navigating Regulations."

As a sole proprietor, all profits will be reflected on Schedule C of your personal tax return. The sole proprietor is responsible for his or her acts and the acts of employees. The sole proprietor's personal assets and business assets are exposed to business risks. If it is a simple enterprise (no employees) with few liabilities (or limited liability) that may be insured against, this entity may meet your needs.

General partnership

A partnership is an association of two or more persons to carry on as co-owners of a business for profit. A general partnership is funded by the amount the partners put in: generally, a partner receives a percentage interest commensurate with the amount of capital (or sweat equity) he or she contributes. The partnership will file an information return; the partner's percentage share will be considered personal income to the partner.

The importance of having a written partnership agreement cannot be overemphasized. Although partnership can be implied from the actions of the partners, such an important fact should not be left to implication and should be set out in a formal agreement. This should address issues of management, additional capitalization, allocation of profits and losses, operational guidelines, dispute resolution, revisions, and termination procedures.

The disadvantage of a partnership is that you are responsible (liable) for your partner. This fact cannot be altered even in a partnership agreement; however, the agreement may contain provisions for contribution and indemnification to cushion the blow.

Limited partnership

Limited partnerships are "layered" partnerships with one or more general partners and one or more limited partners. The general partners run the businesses; the limited partners are passive investors and are prohibited from taking part in management. The liability of limited partners is limited to the amount of their contribution to the partnership.

The name of the entity must contain the words "limited partnership" or "LP," and it must be registered with the California Secretary of State. A limited partnership must obtain a federal tax ID number.

Corporation

A corporation is a taxable entity considered by law to be an artificial person possessing the same rights and responsibilities as an individual. A corporation is formed by filing articles of incorporation with the California Secretary of State. A corporation is also required to have bylaws, which are the rules and guidelines for how the corporation will be run. The corporation will also be required to file an application for an employer ID number with the Internal Revenue Service. The corporation will receive an annual report from the California Secretary of State, which must be filled out and returned to avoid dissolution.

A corporation is liable for its own debts. Its shareholders have limited liability only up to the amount of their investment, so long as the corporation is adequately capitalized and the corporation formalities are observed.

Corporations are subject to double taxation unless the corporation files a "Sub-S" election with the IRS. An S corporation is not subject to double taxation; its income is passed through its shareholders. The election was created for the benefit of small business endeavors. It can, therefore, have no more than seventy-five shareholders who must be individuals (not other corporations) and U.S. residents.

Limited liability company

This hybrid entity offers the limited liability of a corporation, is not double taxed, and is governed by a formal written agreement. It is formed by filing a certificate with the California Secretary of State. The name of the company must contain the words "limited liability company," "limited liability co.," or "LLC." The owners must identify whether it will be managed by the members or by a manager, who may also be a member.

Like a corporation, the limited liability company is liable for its own debts; members and managers are not liable for the debts of the company and are not liable to each other except for gross negligence, intentional misconduct, or a knowing violation of the law.

The LLC is funded by the amount the members contribute for membership interest. The LLC may raise additional capital through its existing members and, subject to the agreement, by admitting new members.

An LLC may be expensive to form, depending on the complexity of the agreement. If the business is managed and the profits divided in a simple standard format, and if the shareholders qualify, an S corporation may be the more practical entity.

Limited liability partnership

General partnerships and limited partnerships can apply to be a limited liability partnership by filing a certificate with the California Secretary of State. A partner in an LLP is liable to the same extent as a partner in a general partnership, except that a partner is not liable for the negligence or malpractice of another partner. If the LLP provides professional services, the LLP is required to carry a specified amount of malpractice insurance.

Identify Your Insurance Needs

In your business plan, make sure you identify the types of insurance your business will require. Purchase whatever insurance you need! See chapter 5, "Developing Your Risk Management Plan," for more information.

Staffing

Compensation rates and reporting requirements of your new enterprise may differ from that of your other farm or ranch employees. Refer to chapter 5, "Developing Your Risk Management Plan." Additionally, make sure you list key employees and their qualifications in the appendix of your business plan.

Meanwhile, in your plan's management strategy, identify the number of employees your enterprise needs and list their titles, duties, and skills. You should also consider the following questions.

- Can I add my new enterprise without overloading my existing staff?
- Do current employees have the right skills to operate this new enterprise?
- How will I recruit additional staff?
- What process will I use to screen and hire employees?
- What training will I provide, by whom, and at what cost?

fies your current and future fixed assets, start-up costs, and several basic forecasts as well as your monthly principal and interest payments.

The financial strategy is fundamental to a lender's evaluation of your enterprise and is key to obtaining money. Make sure to verify your numbers, justify your needs, and accurately research your sources of capital. The financial strategy is equally fundamental to your own evaluation when you address whether you need outside financing.

It is essential that you understand your financial situation. You must know how your financial statements are developed and be able to read and analyze them. Adopt record keeping and book-keeping systems that allow you quick access to accurate data, and make sure that your management system incorporates checks and balances.

Many local community colleges and state universities offer classes to help entrepreneurs set up business books on a home computer. Local Small Business Development Centers offer low-cost classes on how to write business plans and develop financial statements. They also provide no-cost individual counseling to small business owners. Some University of California Cooperative Extension offices offer agriculture business planning workshops. Appropriate Technology Transfer for Rural Areas (ATTRA) has a number of useful factsheets for evaluating enterprises plus agricultural business planning templates on their Web site, http//:www.attra.ncat.org. You can also pay a firm to write your business or marketing plan for you. Check the Internet for firms that will do this. Although you know your business best, if you are averse to writing things down, this service might be helpful. See chapter 7, "Resources for Success," for more information.

- How will I set salaries and wages?
- What benefits will I provide?
- How will my business run if a key person gets ill or injured?

Working with Regulatory Agencies

When you run an agritourism or nature tourism operation, you'll find that you will need to comply with rules that federal, state, and local governments impose in the form of regulations. A little homework is required to complete this part of your business plan. You can't know which regulations affect your business until you've thought through your plan, and you can't finish your business plan until you know what you're allowed to do and the financial cost. See chapter 4, "Understanding and Navigating Regulations."

Form a Marketing Strategy

Your marketing strategy explains what you're selling (goods and services); who you're selling to (your target audience); why you're selling it (industry research and market analysis); and how you're selling it (price, promotion, position, advertising, public relations, placement, and distribution). See chapter 6, "Designing Your Marketing Strategy."

Devise Your Financial Strategy

Your financial strategy identifies your sources of existing debt and your financing needs. It speci-

Make a Budget

A budget is a projection of income and expenses. It is your view of the future, based on your assumptions. When you make a budget, write down your assumptions and evaluate them later for accuracy. Lenders will review them, and you'll use them in next year's budget.

You can make a budget for the entire farm or

Can Agritourism Create a Steady Income Stream?

California agritourism operators responded to a survey mailed in 2009 about their experiences and agritourism business for their 2008 season (Rilla et al. 2011).

Based on the USDA definition of a small farm as having "annual gross revenues of $250,000 or less," 68 percent of the California agritourism survey respondents fit that category. On the other hand, 14 percent of the operators had annual revenues of $1,000,000 or more. Almost half (48 percent) of the operators reported less than $10,000 in gross revenues from their agritourism operations in 2008, while 21 percent had agritourism revenues of $100,000 or more. Respondents with a business plan tended to generate more income than those without one.

Over 50 percent of survey respondents making more than $50,000 found their venues to be profitable. Pumpkin patches and on-farm sales of products were the most profitable types of activities for these operators. There were equal numbers of operators with revenue less than $1,000 as with revenue more than $100,000, and 44 percent percent of small farm operations earned $25,000 or more in agritourism income, which could account for 10 percent of the farm's total income.

Survey respondents indicated a desire for business planning to improve business success. Marketing and management assistance to improve fee revenues for activities currently provided gratis and assistance with effective promotion could also increase the bottom line.

for a single enterprise within the farm. The latter—called an "enterprise budget"—considers only the items of income and expense attributable directly to your new enterprise. When you apply for a loan, you must present an enterprise budget that is integrated within the entire farm budget and a personal budget. Check your local Small Business Development Center for more information about budgets.

Sample budget

Remember Working Landscapes Ranch, the fictional third-generation cattle ranch with the nature tourism mission? Figure 3.5 is a sample enterprise budget for that operation. Note that you will need to use figures from your operation to have a realistic sense of your fiscal situation.

In the budget, the break-even is the projected (actual) revenue (called income on the sample) that must cover the projected (actual) expenses. Gross margin is a financial measure of the efficiency or production of the enterprise without the confusion of business size or business structure. Gross margin is the contribution that each enterprise makes toward the overhead costs. To determine the gross margin, subtract direct expenses, which are costs that change as the units of production change for each enterprise, from gross product or income, which is the value of production. The overhead includes costs that do not change as units of production change (all land and most labor costs are overhead). Gross profit or loss is gross margin minus overhead costs.

Fixed Assets

Fixed assets are the things you own that have a useful life of more than one year. Examples include land, buildings and improvements, machinery, other equipment, livestock, office furniture, and computers. You should determine the cost, estimated acquisition date, and the useful life of each fixed asset, and regularly complete a depreciation schedule. You can also develop a list of assets needed for future development and get special loans only to finance equipment.

Start-Up Costs

Every new enterprise has start-up costs, and you must include these in your financial plan. Start-up costs are the costs associated with opening your enterprise. Most are one-time expenditures, while a few occur every year. Legal fees, accounting fees, licenses and permits, remodeling work, advertising, promotions, and hiring costs are examples of start-up costs.

Sales Forecasts

Sales forecasts are vital from the perspective of both management and sales: you can plan financially only when you have an estimate of sales. Consequently, you need to review your data about products, customers, competitors, and budgets to establish trends and projections for your financial strategy. Be realistic with these forecasts. Based on your research, estimate the total market size in dollar sales per year and create a market share analysis with worst case, most likely case, and best case scenarios. Decide which services you'll provide for free and which ones you'll charge for. If you don't charge for some of the services, how do you expect to generate money? Income could be generated from having something for visitors to see, do, or buy.

Cash-Flow Forecast

When does money come in and when does it leave your operation? The cash-flow forecast is a prediction of your income and expenses as they "flow" through your operation, usually over a period of one year. Many agricultural enterprises have expenses year-round yet receive payment for their product only once or twice during the year. The cash-flow projection allows you to closely estimate how much working capital your new enterprise will require. Adequate working capital is a key requirement for any new venture if the family budget or existing business budget is not to be adversely impacted.

Develop a monthly chart of operations and combine it with your start-up costs, sales forecast, and debt servicing to create a cash-flow budget. Lenders will examine this document carefully.

Sample Cash-flow Forecast

Figure 3.6 is a cash-flow forecast for Working Landscapes Ranch. Chapter 7, "Resources for Success," lists where to find cash-flow forecasts and other business forms online. Also see the book *Writing Business Plans That Get Results: A Step-by-Step Guide*, by Michael O'Donnell (Contemporary Books, 1991).

Profit and Loss Statement

A profit and loss statement shows your profit as a positive result and loss as a negative result.

Profit = (Revenues + Adjustments to revenues) – (Expenses + Adjustments to expenses)

A profit and loss statement can be considered a condensed budget with extras. Unlike an enterprise budget, however, it includes adjustment categories such as capital gains or losses, or depreciation. Sometimes called an income statement, it shows what happens to your business over a specific time period, such as monthly, quarterly, or yearly.

Just as with the enterprise budget, you can create a profit and loss statement for the entire farm or for a single enterprise within your farm, combining its forecast for the existing business. The latter allows you to assess the impact of the new enterprise on the whole. When preparing this statement, remember to accurately account for the start-up phase of your business; it takes time to "get up to speed." Lenders expect to see both actual and projected profit and loss statements.

Sample profit and loss statement

Figure 3.7 shows a profit and loss statement for Working Landscapes Ranch.

Balance Sheet

In a balance sheet, the two sides of the following equation form a balance.

Assets = Liabilities + Owner equity

Figure 3.5

First Year Budget for Working Landscapes Ranch

	Land lease	Cross-country skiing	Enterprise 3	Enterprise 4	Total
INCOME	**26,614**	**13,760**			**40,374**
DIRECT EXPENSES					
Stream restoration	17,068				17,068
Portable paddock fence	500				500
Rehab line shack		1,550			1,550
Hired labor	1,680	2,133			3,813
Marketing		1,245			1,245
Contracts and custom work					0
Equipment rentals					0
Other supplies and services		675			675
Pickup truck costs	144	144			288
Total direct expenses	**19,392**	**5,747**			**25,139**
GROSS MARGIN	**7,222**	**8,013**			**15,235**

Figure 3.5 adapted from Rosenzweig, M.A. 1999. *Market Farm Forms: Spreadsheet Templates for Planning and Organizing Information on Diversified Market Farms*. Auburn, CA: Full Circle Organic Farm.

Assets are what your business owns. Liabilities are what the business owes. Owner equity is what the business owner owns—free and clear of debt—at a specific time.

You must include a balance sheet in your business plan. Because you're dealing with assets and liabilities, the balance sheet will probably have to involve your entire operation. You can evaluate your new enterprise's impact on the entire operation, however. To do so, produce one balance sheet that includes the new enterprise's assets and another that excludes them. Your accountant can prepare your balance sheets or can show you how to prepare your own; or, see the Business Owner's Toolkit Web site, http://www.toolkit.cch.com/. For a sample balance sheet, see figure 3.8, which is the balance sheet for Working Landscapes Ranch beginning at Year 1.

Other Financial Documents

From the documents you've developed, you can calculate a projected income statement, a break-even analysis, and other necessary financial information such as sources and uses of cash. These are excellent tools for helping you to assess the feasibility of your new enterprise.

Write a Financial-Strategy Conclusion

Your financial strategy should include a conclusion that summarizes your financial documents and explains how your new enterprise will fit into your current operation. It is the second part of your executive summary, the first part being the business concept.

Include an Appendix

The appendix is located at the very end of your business plan. In it, you can include supporting documents such as

- tax returns
- major financial documents, both business and personal, including prior credit and debt history
- your résumé and those of key employees
- résumés of your management team, including your outside consultants and advisors such as accountants, lawyers, bankers, and insurance brokers
- certificates of completion of all business planning and management courses you've taken

Find Financing

If you're like most small-business owners, you'll require capital to start your agritourism or nature tourism enterprise. How do you obtain money?

There are a number of sources, including personal funds, debt financing, equity financing, traditional agricultural lenders, small farm–friendly banks and holding companies, and the U.S. Small Business Administration (SBA), as well as rural economic development agencies, councils, and districts. Using any of these entails risk—borrowing inherently brings risk, either financial or personal.

Personal Funds

Personal funds include

- savings
- spouse's income
- cash from second mortgages
- gifts from family and friends
- cash from selling part of the property or an easement
- cash from credit cards

Debt Financing and Equity Financing

Debt financing is borrowing money or taking on debt to further your business. You still own your business, and you still make all of your business decisions. Equity financing is selling a piece of your business. You no longer own the entire business, and you are usually accountable to other people when making decisions.

Equity financing includes limited partnerships or stock offerings, both of which require professional legal advice and assistance.

Traditional Agriculture Lenders

Traditional farm lenders include the Farm Credit Service and the Farm Service Agency. The Farm Credit Service is a collection of federally chartered borrower-owned credit cooperatives, the Farm Credit Cooperative Banks. They lend to agricultural operations and provide rural home loans.

Another traditional agriculture lender is the Farm Service Agency, part of the U.S. Department of Agriculture. It has a direct lending program and a loan-guarantee program, providing money when other lenders won't.

Figure 3.6

FY combined cash flow (land lease and cross-country enterprises) for Working Landscapes Ranch

	Mar	Apr	May	Jun	Jul	Aug	Sep	Oct	Nov	Dec	Jan	Feb	Total
CASH INFLOW													
Land lease		3,014	3,013	3,013	3,013	3,013	3,014						18,080
Cross-country ski access									1,835	3,669	4,587	3,669	13,760
WHIP cost-sharing monies										8,534			8,534
Total inflow		**3,014**	**3,013**	**3,013**	**3,013**	**3,013**	**3,014**		**1,835**	**12,203**	**4,587**	**3,669**	**40,374**
CASH OUTFLOW													
Stream rehab for grazing, etc.	10,419		6,325	325									17,069
Portable electric fence		500											500
Hired labor	840		840						533	533	533	533	3,812
Rehab line shack to warming shed				1,500									1,500
Plot and reproduce trail maps									50				50
Marketing for cross-country				1,085	20	20	20	20	20	20	20	20	1,245
Additional liability insurance								475					475
2 hours' attorney time				200									200
Truck costs	24	24	24	24	24	24	24	24	24	24	24	24	288
Total direct costs	**11,283**	**524**	**7,189**	**3,134**	**44**	**44**	**44**	**519**	**627**	**577**	**577**	**577**	**25,139**
Interest on opportunity capital	33	113	113	313	113	113	113	113	113	113	33		1,283
Total opportunity cost	**11,316**	**637**	**7,302**	**3,447**	**157**	**157**	**157**	**632**	**740**	**690**	**610**	**577**	**26,422**
OVERHEAD COSTS													
Office expense	50	50	50	50	50	50	50	50	50	50	50	50	600
Farm insurance	200		100	100	100	100	100	100	100	100			1,000
Property insurance	400												400
Property taxes		429									429		858
Maintenance	44	44	44	44	44	44	44	44	44	44	44	44	528
Total cash overhead	**694**	**523**	**194**	**194**	**194**	**194**	**194**	**194**	**194**	**623**	**94**	**94**	**3,386**
TOTAL CASH COSTS	**$12,010**	**$1,160**	**$7,496**	**$3,641**	**$351**	**$351**	**$351**	**$826**	**$934**	**$1,313**	**$704**	**$671**	**$29,808**
COMBINED CASH FLOW	($12,010)	$1,854	($4,483)	($628)	$2,662	$2,662	$2,663	($826)	$901	$10,890	$3,883	$2,998	$10,566

Combined cash flow = total cash inflow minus total cash costs. *Source:* Adapted from Rosenzweig 1999.

An October 2009 USDA publication, Building Sustainable Farms, Ranches and Communities, has a comprehensive listing of federal resources and programs.

Also see the publication Strategies for Financing Beginning Farmers (Center for Rural Affairs, http://www.cfra.org/files/BF-Financing-Strategies.

pdf) and the National Council of State Agricultural Finance Associations Web site (http://www.stateagfinance.org/types.html).

Small Business Administration

The SBA offers two primary loan programs that provide funding to small businesses unable to obtain capital through normal lending channels.

Figure 3.7

Projected profit and loss statement for Working Landscapes Ranch for period from March 1, 2002, to February 28, 2003

INCOME (from sales)	Budget		Receivable	+	Cash
Land lease: cattle	18,080				18,080
Cross-country skiing	13,760				13,760
Enterprise 3					
Enterprise 4					
Enterprise 5					
Enterprise 6					
Enterprise 7					
Enterprise 8					
Other farm income (subsidies)	8,534				8,534
Total income			40,374		
EXPENSES	**Budget**		**Payable**	**+**	**Cash**
Stream restoration project	17,069				17,068
Portable electric fence	500				500
Rehab line shack	1,550				1,550
Hired labor	3,812				3,813
Marketing	1,245				1,245
Contracts and custom work					
Other supplies and services	675				675
Pickup truck costs	288				288
Building repairs and maintenance	528				528
Equipment repairs and maintenance					
Land rent (mortgage)					
Land and water taxes	858				858
Legal and accounting fees					
Insurance	1,400				1,400
Other office	600				600
Miscellaneous					
Operating interest (8%)	1,283				1,283
Term loan interest					
Total expenses			29,808		
EXCESS INCOME OVER CASH EXPENSES			10,566		
Adjustments					
Less depreciation			15,826		
+ Ending inventory					
– Beginning inventory					
= Inventory change (+ or –)					
NET FARM INCOME			**–5,260**		

Note: For income tax purposes, all improvements made in the first year were expensed.
Source: Adapted from Rosenzweig 1999.

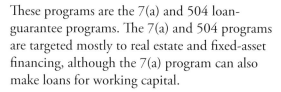

These programs are the 7(a) and 504 loan-guarantee programs. The 7(a) and 504 programs are targeted mostly to real estate and fixed-asset financing, although the 7(a) program can also make loans for working capital.

SBA loans are made through private lenders. Many rural and small-town banks are able to make SBA-guaranteed loans. The SBA itself has no funds for direct lending or grants.

The SBA has recently established an effective microloan program that makes loans of up to $35,000 through designated intermediary lenders.

This program is available in many rural areas.

More information on these loan programs is available from your lender, your local SBA office (see the SBA Web site, http:// www.sba.gov), and the Small Business Development Center. Also refer to chapter 7, "Resources for Success."

Rural Economic Development Agencies, Councils, and Districts

Rural economic development agencies oversee, distribute, and lend monies from federal and state community-development block grants and

Figure 3.8

Balance sheet for Working Landscapes Ranch as of March 1, 2002

ASSETS				LIABILITIES		
CURRENT	Cost	Market value		**CURRENT**	Cost	Market value
Cash on hand				Operating loan	15,000	15,000
Stocks, bonds, savings				Accounts payable		
Accounts receivable	8,534	8,534		Taxes payable	858	858
Crops inventory				Rent payable		
Crop supplies				Loan principal due within 1 year		
Prepaid expenses				Intermediate loans		
Nonfarm and personal	5,000	5,000		Long-term loans		
Other				Other insurance	600	600
Total current	**13,534**	**13,534**		**Total current**	**16,458**	**16,458**
INTERMEDIATE (1–10 yrs)				**INTERMEDIATE (1–10 yrs)**		
Machinery and equipment				Notes payable		
Cost (basis)	38,215			Machinery and equipment		
Less accumulated depreciation		38,215		Vehicles		
Nonfarm and personal				Contingent tax liability		
Other				Other		
Total intermediate	**38,215**	**38,215**		**Total intermediate**	**0**	**0**
FIXED				**LONG-TERM (>10 years)**		
Buildings				Building loans		
Cost (basis)	230,000			Land mortages		
Less accumulated depreciation		230,000		Contingent tax liability		
Land	316,094	316,094		Other		
Nonfarm and personal				**Total long-term**	**0**	**0**
Other				**TOTAL LIABILITIES**	16,458	16,458
Total fixed	**546,094**	**546,094**		**OWNER'S EQUITY**		
				+Beginning equity	581,385	581,385
				+Net farm income		
				+Owner's contributions		
				–Owner's withdrawals		
				= Ending equity	**581,385**	**581,385**
TOTAL ASSETS	**597,843**	**597,843**		**TOTAL LIABILITIES AND OWNER'S EQUITY**	**597,843**	**597,843**

Source: Adapted from Rosenzweig 1999.

from USDA Rural Development Agency lending programs. Their funds are often tagged for specific purposes such as job creation or retention, housing rehabilitation, rural infrastructure tied to increased employment, microloans to start-up businesses with job creation potential, and rehabilitation of community facilities.

Threats to Success

Despite the optimism that surrounds the start-up of a small business, most small businesses fail: 80 percent of small businesses fold in their first five years. However, remember that most large businesses started out as small businesses. Remember too that understanding threats to success can help you avoid problems. Unquestionably, all entrepreneurs make mistakes, but those who recognize and correct their mistakes are more likely to succeed.

Why do small businesses fail? The answer depends on their owners' entrepreneurial skills and overall business conditions. It is important to have good analytical, communication, and interpersonal skills in addition to drive and vision. It is also important to understand common threats to business success. The ensuing list outlines some recurring pitfalls (Ward 1997).

Failure to understand or predict the difficulty of owning your own business or adding an enterprise. Owning and running a business is hard and stressful work. As a service business operated from your home on family land, your agritourism or nature tourism enterprise can be particularly taxing,

requiring tremendous commitment and effort. Learn how to cope with stress. Have balance in your life; enjoy family, friends, and activities that give life meaning.

Lack of strategic planning. Make mistakes on paper before you risk assets! Understand that planning is an ongoing and cyclical process. In other words, plan strategically. Strategic planning consists of

- developing a clear mission statement
- assessing your company's strength and weaknesses
- analyzing your market
- analyzing your competitors
- setting goals and objectives
- formulating strategic options and selecting appropriate strategies
- translating your plans into actions
- establishing accurate controls
- evaluating risks, rewards, and feasibility

Not knowing how to manage and operate a business. Agritourism and nature tourism enterprises require a different set of skills than production agriculture. If you don't know what you're doing, learn. Take courses, read books, attend seminars, listen to tapes, get a mentor. Go into business with your eyes wide open.

Lack of cash; poor cash flow; lack of financial savvy; poor money management; underfinancing.

"Do not have the build-it-and-they will-come mentality!" said one California agritourism operator. Instead, know and fill demand—and then expend capital. Learn about available financing opportunities, tools, and techniques. Understand the basics of business finance. Think of capital as the gas in your gas tank. That is, have enough gas to reach your destination, because you can't count on finding a gas station along the way, and if you do it may not be affordable!

Growing too fast. When a business grows too fast, entrepreneurs lose control, the quality of service goes down, and clients do not appear. Routinely reassess your enterprise, keeping only competent employees and profitable, relatively trouble-free, enjoyable services.

Poor interpersonal skills. Relationships within a business are complex and worthy of time, attention, and feedback. Learn and follow good management rules about hiring, training, growing, and developing employees. Pay employees well. Pay attention to their basic needs. Don't overburden them. Compliment them. Be flexible and deal directly with conflict. Take the time and effort to build a team!

Poor communication skills. Good communication skills consist of good listening as well as good talking. You must express your feelings as well as ideas and be able to apologize if the need arises. Provide a role model for the people who work with and for you.

Failure to innovate. Businesses have been lost because another company jumped into the market and ran with it. Never be complacent!

Trying to go it alone. Businesses can sink when the operator wears too many hats. Having strong, well-rounded management and working teams is central to success. Pay for and use the services of professionals! Hire employees who complement your weaknesses and delegate responsibility to them. Reach out to local residents. You should also form networks with people involved in agritourism and nature tourism activities, both locally and regionally.

Failure to recognize your own strengths and weaknesses. The better you know yourself, the better your chance of avoiding problems and of nurturing your company. Ask for help when you need it.

Failure to seek and respond to criticism. Many problems can be corrected or avoided with the help of candid feedback. Don't blame external causes for failures; take a look at yourself instead. Ask people you trust to critique you. Talk with board members, old friends, consultants, and peers—and do not get angry over their responses! Instead, learn and improve.

Points to Remember

- A business plan is critical to your operation's success. Remember, failure to plan is planning to fail.

- The business plan contains many components to be considered and written: executive summary, mission statement, business concept or idea, measurable goals and objectives, background information (industry research and market analysis), management needs and management history, marketing strategy, financial strategy, and appendix.

- Creating a business plan allows you to anticipate your new enterprise's opportunities and challenges on paper—before you commit substantial resources.

- A business plan is essential if you want outside financing.

- A business plan may cover your entire operation or a single enterprise within it.

- To predict how your new enterprise will impact your entire operation, compare a financial statement of your new enterprise with a financial statement without it.

- There are affordable funding sources available for agricultural expansion into tourism, and there are people available to help you with the funding process.

Acknowledgment

The section "Identify Your Legal Structure" was excerpted from the Small Business Start-Up Information Package and Regional Resource Guide (Spokane, WA: U.S. Small Business Administration, 1991).

Chapter 4

Understanding and Navigating Regulations

Chapter Goals

This chapter will help the landowner

- become familiar with federal, state, and local regulations that affect agritourism and nature tourism enterprises

- anticipate the permits and licenses required for these enterprises

- prepare for the permitting process

- estimate the costs associated with permits and licenses

- communicate effectively with agency staff

- move through the regulatory maze

Understanding Regulations

Agritourism supports local farms and ranches and surrounding communities by generating revenue, but diversifying comes with challenges. Zoning, permitting, and environmental health regulations are the leading impediments to farmers and ranchers who want to expand their operations to include agritourism enterprises. Comments from respondents to the 2009 state agritourism survey echo their frustration (Rilla et al. 2011). Regardless of the region or county in California, they are confused and overwhelmed with their county policies, procedures, and the related expenses involved in initiating or expanding an agritourism enterprise. You can best address the regulatory bureaucracy by taking it one step at a time. This chapter helps you understand these regulations and know where to start with them when creating and operating your agritourism or nature tourism enterprise.

Regulations are legal requirements imposed by federal, state, and local governments. These regulations go hand-in-hand with developing a business plan: you can't know which regulations affect your business until you've thought through your plan, and you can't finish your business plan until you know what you're allowed to do and the financial cost of doing it.

Introducing agritourism activities to your facility triggers complying with additional legal requirements. If you are fresh to agricultural production and processing and you are also starting an agritourism enterprise, you have a lot to learn about regulations!

California law consists of twenty-nine codes that cover a variety of subjects. As you contemplate a new enterprise, you may be interested in the Fish and Game Code, the Health and Safety Code, the Food and Agriculture Code, the Business and Professions Code, and the Labor Code. The text of these codes can be found at the Official California Legislative Information Web site, http://www.leginfo.ca.gov/calaw.html. Many county codes are also available at the Municode Library Web site,

The more activities you add, the more regulations you invoke. Depending on your plans, meeting the regulatory requirements might be as simple as limiting the size of your produce stand to the maximum size allowed in your zone. Or it might be as complex and costly as installing a new septic system, a handicapped-accessible bathroom, and a commercial kitchen to your on-farm cafe.

> "Ultimately, we're trying to decide: Do applicants need a food permit? Do they need to do something to a septic system—put in a new one or is the existing one okay? Do they need a public water supply? These are the three main categories we deal in. Plus, are they going to have a noise impact to the surrounding community?"
>
> *County Environmental Health Specialist*

http://www.ordlink.com. Note that reading the code will not give you a complete understanding of the law, however, since case law and administrative decisions play a role in how the code is understood and applied. Readers should contact the appropriate agency for information and consult their attorney.

If you are a farmer or rancher, you are already familiar with the regulations impacting your operation. If you own a postharvest or processing operation as well, you are acquainted with those regulations too. Therefore, it will probably be no surprise to you to learn that introducing agritourism or nature tourism activities to your facility creates additional legal requirements.

But what if you're new to agriculture? If you're fresh to agricultural production and processing and are also starting an agritourism or nature tourism enterprise, there are excellent resources to help you on your way.

Looking at Basic Regulation Categories

To start an agritourism enterprise, you must comply with local regulations, and you may need to obtain county zoning approval. You must adhere to regulations that protect the environment, your neighbors, and your visitors. The following sections describe the categories of regulations relevant to agritourism.

Understanding Land Use and Land Development Regulations

County Zoning

Jurisdiction: County
Contact: Planning department

Complying with zoning is the first step in establishing your agritourism enterprise. Contact your county planning department about zoning laws that pertain to your property.

Each county has a general plan that describes its land use policies. Maps that show the county's land use zones are usually included in the general plan. County zoning governs how parcels in each land use can be used. Typically, counties update their general plans every ten years. This update requires a review and sometimes a revision of the zoning and development codes. It's a good idea to stay familiar with the zoning and development codes that apply to your land.

Agricultural zoning policies protect farmlands and ranch lands from uses detrimental to agriculture. Zoning codes list allowed farming uses and compatible nonfarming uses for agriculturally zoned land. Each county decides what uses to allow and what conditions or standards to impose on these uses.

Zoning codes distinguish between nonfarm uses that are permitted by right and uses that require a use permit. Uses that require a use permit are subject to conditions, review, and approval. If a land use is permitted by right, it is allowed without special approval as long as it meets zoning and other requirements.

Each county defines its own permit types. Many counties have more than one type of use permit, each with its own type of requirements. The section "Putting the Pieces Together: The Permit Approval Process" in the online ANR publication *Agritourism Enterprises on Your Farm or Ranch: Where to Start* (George 2008) explains the steps you must follow to obtain a use permit.

Some counties have special ordinances for certain agritourism enterprises. These ordinances specify legal requirements and often offer a streamlined permitting process.

Produce Stands

Some counties do not require a permit for produce stands, but they still require owners to complete a form and pay a fee. The process usually requires a copy of the deed or a legal description of the property, a written description of the use requested, and a detailed sketch showing the location of existing structures with respect to road intersections, existing buildings, and signs. Check with your county planning department.

Roadside Farmstands and U-Pick

Typically, roadside farmstands and U-pick farms are permitted by right in agricultural zones. If, however, you plan to offer recreational activities or overnight accommodations on your property, to hold public events, or to process and serve food to the public, you probably will need a use permit. Requirements for that permit could include meeting building, health, and safety codes; improving septic conditions, parking, and public road access; or other public safety rules. See the section "Understanding Public Health and Safety Regulations" in this chapter.

Building Codes

Jurisdiction: County

Contact: Building department

If you make structural changes to your operation, the changes must conform to building code standards. If you renovate old buildings or construct new facilities, you must obtain a building permit from your county's building codes office. Note that all public-use structures must conform to accessibility standards of the Americans with Disabilities Act.

Roads and Traffic

Jurisdiction: State and county

Contact: CALTRANS, county planning department, and county public works department

Your agritourism enterprise means that visitors will come to your farm or ranch, which means increased traffic on the roads. If your proposal requires a use permit, your county's planning and public works departments or CALTRANS (or both) will review your proposal for public

safety concerns related to roads and traffic. They might require that you widen or grade the road passing your property or build a turn-off or bridge.

Use of Public Lands

Jurisdiction: Federal or state

Contact: District office of the Bureau of Land Management or other appropriate agency

If you conduct commercial activities or business on federal- or state-owned lands, you must obtain a special-use permit from the land management agency on which the activities are proposed. Permits are required for certain recreational activities that host or charge fees when the activities take place on public lands. These activities could be an ongoing part of your

County Agencies

Planning Department: Oversees and makes recommendations on land use issues affecting county lands and coordinates the approval process for land use permits.

Building Department: Enforces building codes for new or remodeled structures and issues building permits.

Environmental Health Department: Enforces health codes for food facilities and evaluates new development plans for adequacy of public water supply, septic systems, and environmental impacts of waste disposal.

Public Works Department: Analyzes traffic impacts of development projects and maintains county roads. Issues permits to post signs on county roads.

Agricultural Commissioner, Weights and Measures Division: Oversees the organic registration process, which can include a site inspection and the use of organic labels. Regulates commercial weighing and measuring devices to ensure accuracy. Issues farmers' market permits. Issues certified producer certificates that allow farmers to sell at certified farmers' markets.

Fire Agency: Establishes fire prevention codes for the county and enforces them through safety and occupancy inspections.

Tax Collector: Collects taxes for taxing agencies within the county and issues business licenses.

SIERRA VALLEY FARMS
1329 COUNTY ROAD A23
530 832-0114

operation such as horseback riding, guided fishing or hunting activities, or events such as endurance rides, eco-races, chuckwagon dinners, and so on.

The purpose of the permit process is to

- identify any potential land use or resource conflicts that might arise

- identify applicable procedures, permits, and special conditions needed to protect resources and public uses

- achieve common understanding between the agency and the applicant about the proposed uses

- clarify timeframes, limitations, and responsibilities

It is important that you make early contact (a year in advance isn't too early) with your local representative of the Bureau of Land Management or other appropriate agency to discuss your proposal. District personnel will help you determine whether the proposal justifies submitting a special-use permit application. If you file an application for a special-use permit, you must provide a map and written description, detailed and specific information about your event or activity, and proof of insurance (with the agency listed as additionally insured).

Then, based on your permit application, project description, and potential environmental impacts, the district ranger makes the decision whether to issue the special-use permit. The capacity of the proposed activity and interest from additional applicants may result in a competitive process for issuing permits. You must pay a processing fee and perhaps a monitoring fee before you receive your special-use permit. Some agencies also require that you pay them a percentage of the fees you collect. Special-use permits can be for one-time, one-day events (such as a bike and run eco-race) or for up

to five years, covering activities such as horseback trail rides in conjunction with your ranch.

Signs

Jurisdiction: State or county
Contact: CALTRANS, county planning department, and public works department

Signs identifying and advertising your enterprise are subject to regulations. Signs on your own property are subject to county development and design codes; signs on county roads are subject to county public works department regulations; and signs on state highways are subject to state transportation department regulations.

Understanding Public Health and Safety Regulations

Food Safety

Jurisdiction: State law
Contact: State and county environmental health departments

The California Uniform Retail Food Facilities Law (CURFFL) is part of the California Health and Safety Code. This state law governs food sanitation for food handling, processing, and preparation activities of the agritourism industry.

Specifically, CURFFL details food-safety requirements concerning food handling, equipment, and storage. These requirements are generally stricter for processed foods than they are for nonprocessed foods. According to CURFFL, processed foods sold to the public cannot be prepared in a home kitchen.

CURFFL is enforced at the local level by the county environmental health department. This department reviews plans and regularly inspects food facilities in accordance with state law. If you plan to construct a food facility (for example, to sell pies made from fruit grown on your farm)

California State Agencies

Alcoholic Beverage Control Department: Issues liquor licenses.

CALTRANS: Reviews development proposals for traffic-flow impacts; issues permits to post signs on state highways.

Department of Industrial Relations: Sets occupational health and safety standards (employers must have a health and safety plan).

Public Health Department (in the Health and Human Services Agency): Enforces the California Health and Safety Code and inspects food processing facilities regarding products exported from the county.

Note: The California Environmental Quality Act (CEQA) establishes environmental protection standards, enforced at the county level.

Federal Agencies

U.S. Forest Service Bureau of Land Management (BLM): Issues permits for commercial use of public lands.

Internal Revenue Service: Issues Employer Identification Number (EIN) to employer.

What's the Difference between a Use Permit and a Business License?

Use permits apply to your property and are generally issued when your business is started. Business licenses apply to the operator. They are issued to the operator, usually renewed annually, and transferable to a new business location if the operation moves.

Explained one farmstay operator on California's Central Coast, "A use permit is a one-time fee, and it goes with the land. If we sell the land and go someplace else, the use permit is still okay for whoever buys this ground, and they can do a farmstay on it. But if we move to another place, we have to get another use permit for that area. With a license, you're licensed to do something, and that goes with you."

or modify an existing structure into such a facility, the county environmental health department first must approve your plan. Log onto your county's environmental health department Web site for CURFFL policies or see the California Department of Health Services Web site, http://www.dhs.ca.gov/, and search the site for CURFFL. The text of the law can be found in the California Health and Safety Code §§ 113700 to 114475.

Food Handler's Certificate

Jurisdiction: State and county
Contact: County environmental health department

CURFFL mandates that at least one person in an enterprise processing or preparing food for the public possess a current food handler's certificate. This person may be either the business owner or an employee. You or your employees can earn a food handler's certificate by taking food safety courses from local service agencies. Contact your county environmental health department to learn when and where courses are offered.

Agricultural Homestays

Jurisdiction: State and county
Contact: County environmental health department

Some agricultural operators establish on-farm bed and breakfasts (B&Bs), also called farmstays or agricultural homestays. A 1999 law, the California Agricultural Homestay Bill (AB 1258), amended California's Health and Safety Code to include an agricultural homestay establishment. This law allows working farms to host a limited number of overnight visitors and permits farm families to serve meals cooked in the farm kitchen to visitors, consistent with the federal food code definition of a family home kitchen.

An agricultural homestay establishment must meet all of the following requirements:

- It contains no more than six guest rooms and hosts no more than fifteen guests.
- It provides overnight visitor accommodations.
- It serves food to registered guests only and serves meals at any time, with those prices included in the price of overnight accommodation.
- Lodging and meals are incidental and not the primary function of the agricultural homestay establishment.
- The agricultural homestay establishment is located on the farm and is part of a farm, as defined, and produces agricultural products as its primary source of income.

Farmstay kitchens are regularly inspected by the county department of environmental health. Agricultural homestays must meet all other applicable state and local regulations and zoning requirements as well. Article 18 of CURFFL addresses the regulations for agricultural homestay establishments in its section "Restricted Food Service Transient Occupancy Establishment."

Farmstand and U-Pick Regulations

3G Family Orchard—the farmstand, bakery, and restaurant enterprise discussed in chapter 3—would have an easy time if it simply sold only raw produce at a farmstand. Why? Most counties consider roadside farmstands and U-picks to be compatible uses in agricultural zones. Accordingly, they require no permit. Consequently, these two activities are among the easiest to set up from a regulatory standpoint.

Some regulations do exist, however, especially for farmstands. Restrictions might include

- a minimum number of acres in permanent agricultural crop production
- requirements specifying a minimum percentage of produce sold to be grown on your farm
- limitations on structure size
- setback from the road
- requirements regarding sign size and location
- provisions for parking

"If you're selling a product that you grew right off the farm, I don't think the county would do much. But once you start putting something in a jar or cooking, then you're going to have to go through the county. And they're going to have some rules and regulations for you. Once our product went out of our county, then it involved the state. So, now, we have the state inspector come. He checks our pie shop and checks our juice room and all of that."

Pie Shop and U-Pick Operator

Public Safety

Jurisdiction: County
Contact: County fire agency, sheriff's department

Agricultural operations involving the public must comply with public safety and accessibility regulations such as those concerning emergency exit access (fire safety) and restroom facilities. There are also local ordinances and codes for fire prevention and safety, including requirements relating to fire extinguishers and on-site water. Contact your local fire agency and sheriff's department to ensure that your activities conform to local ordinances.

Understanding Environmental Health Regulations

Water

Jurisdiction: County
Contact: Environmental health department

The county environmental health department reviews agritourism or nature tourism proposals to make sure there is adequate (in terms of quantity and quality) on-site water for visitors.

Septic Systems

Jurisdiction: County
Contact: Environmental health department

Each county has its own septic standards and usually requires a permit to repair, upgrade, or construct a septic system. If you're starting with a parcel of raw land, you'll need a site evaluation. That evaluation determines the suitability of on-site sewage disposal and the sewage treatment system that best fits your site and soil. If you have an existing septic system, the environmental health department will evaluate your proposed site for its adequacy to accommodate farm visitors.

To avoid excessive cost and aggravation, it is important that you work closely with your county environmental health department. Alternate or additional systems may be required even if they do not actually seem necessary.

Understanding Direct-Marketing Regulations

Jurisdiction: State and county
Contact: County agricultural commissioner

Weights and Measures

If you sell goods directly to the public by weight, California and other states require that you annually license your scales with the county weights and measures division. This requirement ensures that your scales are properly calibrated and have passed inspection.

Package Labeling

If you package your fresh or processed products, you must attach a label that identifies the product, your business name and address, and the net contents in terms of weight or other measure. The agricultural commissioner's weights and measures division enforces this requirement.

Organic Registration

If you produce and sell organic products in California, the California Organic Foods Act of 2003 requires that you register your enterprise with your county agricultural commissioner. In addition, if you produce or sell an organic product and your gross sales are over $5,000 annually, you must also be certified by a USDA-accredited organic certifier.

Farmers' Market Permits

If you operate a booth and sell produce or certifiable agricultural products at a certified farmers' market, you must obtain a Certified Producers Certificate (CPC) from the county agricultural commissioner where the market resides.

Seasonal Processing

If you have seasonal processing needs, consider renting a commercial kitchen from a local church or a school cafeteria rather than building one of your own.

> "We've always had the potties out there. They were called outhouses years ago. So we brought in the plastic potty houses. We kept a water tank out there then, and we'd just wash up or have a drink right out of the water tank. Well, that's all changed. You have to have the regular facilities right on the potty house—attached to it—for washing and drying your hands and everything. Those rules have really changed a lot. For the U-pick, that's one of the biggest added expenses—the potties."
>
> *U-Pick Operator*

Understanding Business Regulations

Fictitious Business Name Statement

Jurisdiction: State law
Contact: County recorder's office

If you select a business name that includes neither your surname nor another owner's name, you must file a fictitious business name statement. You can find and file that form at the office of the county clerk or recorder. Your statement then will become a public document, kept on file until it expires or you abandon your endeavor. To avoid redundancy with other enterprises, search names already on file as you contemplate the name of your new enterprise. Within thirty days of filing the fictitious business name statement, you must publish it in a newspaper of general circulation within your county. You must do so once each week for four successive weeks.

Business License

Jurisdiction: County
Contact: Tax collector's office

Every person who does business in a California county must hold a California business license. In general, business licenses are issued after you've received clearance from other county departments. To obtain your business license, get an application form from the county tax collector's office.

Complete it and submit the application form along with a nonrefundable fee to the county tax collector's office. Your form will go to all county departments for review and approval. Once it's approved, you'll receive a business license in the mail. You must post this license in a conspicuous place on your premises. And you must renew it annually, paying a fee each time you do so.

Liquor License

Jurisdiction: State
Contact: California Department of Alcoholic Beverage Control (ABC)

There are many kinds of liquor licenses. In California, log onto the Web site of the Department of Alcoholic Beverage Control, http://www.abc.ca.gov, for extensive and detailed information and instructions.

Applying for your liquor license requires that you visit the nearest ABC office (see the ABC Web site for office locations). According to ABC instructions, you must apply in person because "considerable detailed personal information is required, including fingerprints from all individual applicants, managers, and managing officers of applicant corporations. Moreover, it is desirable for an employee of the department to advise all applicants in person of pertinent laws, rules, and regulations."

It will take ABC from thirty to fifty days to issue your license. You can apply for a new license or you can have a license transferred to you by a person or business with an existing license. Licenses must be renewed yearly, and they require an annual renewal fee.

Transient Occupancy Tax

Jurisdiction: State and county
Contact: County tax collector

If you operate an agricultural homestay or farm B&B, you must pay a transient occupancy tax (TOT). The TOT is a state tax on revenues generated locally from the hotel, motel, and accommodations industry. In many counties and in some incorporated towns, the TOT includes a local tax as well as the state tax. County TOT rates range

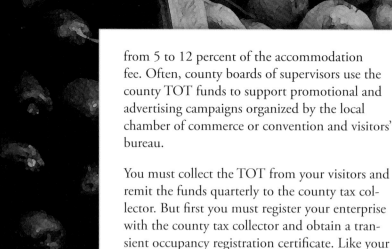

from 5 to 12 percent of the accommodation fee. Often, county boards of supervisors use the county TOT funds to support promotional and advertising campaigns organized by the local chamber of commerce or convention and visitors' bureau.

You must collect the TOT from your visitors and remit the funds quarterly to the county tax collector. But first you must register your enterprise with the county tax collector and obtain a transient occupancy registration certificate. Like your business license, this certificate must be displayed conspicuously on your farm or ranch.

Labor Laws

Labor laws are often the most difficult with which to comply. Because the number of regulations is increasing, you must know the laws governing migrant labor, minimum wages, workplace safety, and special taxes.

The California Chamber of Commerce, http://www.calchamber.com, is a good resource for federal and state labor law information. The U.S. Department of Labor's Office of Small Business Programs helps small businesses comply with rules, regulations, and laws enforced by the U.S. Department of Labor; see their Web site, http://www.dol.gov/dol/osbp, or call toll-free at 888-972-7332. Additionally, the University of California Agricultural Personnel Management Program, http://apmp.berkeley.edu, offers information about labor-related topics.

Employer Identification Number

Jurisdiction: Federal
Contact: Internal Revenue Service

If you hire employees specifically for any of your agritourism enterprises, you must obtain an employer

identification number (EIN) and a federal tax identification number. When you send a completed Form SS-4 to the IRS, you will register your business with the IRS, the Social Security Administration, and the Department of Labor. You will receive a federal tax identification number in the mail.

Occupational Health and Safety

Jurisdiction: State
Contact: Department of Industrial Relations

Businesses that hire employees must prepare an injury and illness plan. To help you create a safe and healthy workplace, the state offers a no-fee consultation. In California, you might have to obtain certain permits, licenses, and certificates to comply with the state's health and safety standards.

Employee State Tax Registration

Jurisdiction: State
Contact: Employment Development Department

Employers must register with the state for employee state-tax issues.

Regulations and Your Business: Where to Start

If you take this regulatory journey one step at a time, you'll find it to be much easier. Research the process, then ask a lot of questions. The more you know in advance, the better questions you'll ask and the better prepared you'll be to prevent costly delays and difficulties.

Do Your Homework

First make sure that your county allows your kind of enterprise to operate in the designated land use zone. Start with a visit to your county planning department. Bring your assessor's parcel number, which can be found on your tax bill. It will help you identify which zone covers your property.

Winegrowers Can Sell at Farmers' Markets

In 2001, a section was added to the California Business and Professions Code allowing winegrowers to sell a prescribed amount of their wine off-sale at certified farmers' markets. The law contains many limitations and regulations, however, which are listed on the Alcoholic Beverage Control (ABC) Web site. Permits must be renewed annually, along with the master winegrower's license.

ABC has developed a simplified application procedure for licensed winegrowers to obtain this permit (Type 79). You can apply for your permit from any district ABC office. But before you apply, ABC recommends that you obtain the required product certificate(s) from the county agricultural commissioner and permission from the certified farmers' market to sell wine at that particular site. The California Federation of Certified Farmers' Markets provides information at their Web site, http://cafarmersmarkets.net/.

Understanding and Navigating Regulations Case Study

Double T Ag Museum

After twenty-seven years of dairy farming, Tony and Carol Azevedo now host weddings. Tony has collected many horse-drawn wagons, carriages, and farm equipment. The museum, opened in 1992, is a tribute to all the hard work that farmers and their families had to do to make a living. Tony and Carol provide educational programs at their museum and host special events. Double T Ag Museum is located near Stevinson, California. They call their operation "agritainment."

It all started with the five-hundred-guest anniversary party for Tony's parents. Soon afterward, the Azevedos were fielding calls to hold additional functions. "We found out that the American public is looking for an adventure as long as the bathrooms are clean and the food is good," Tony says.

So the Azevedos embarked on an opportunity to save the family farm. Neighboring farmers were skeptical. Bankers were dubious. And the Azevedos found out they needed a $1 million liability insurance policy.

Still they persevered, keeping their homeowner's policy because "a farm is a residence." In addition, they required clients to obtain insurance, acquired simply through the person's homeowner's policy. Insurance companies rarely charge for a night's entertainment. And, says Tony, if there was an accident (which there hasn't been), the client's policy pays first. This strategy encourages prudence among guests and protects the Azevedos from lawsuit artists.

Most guests come from outside of the local area, from Turlock, Modesto, and beyond. Many come from the Bay Area—in quest of adventure, speculates Tony. Their guests say they like being off the main road, feeling better able to enjoy their evening with their children safe from traffic and strangers.

Although the Azevedos hold class reunions and company parties, weddings compose 95 percent of their business. Once the clients reserve a date, Carol provides them with a list of items they need to bring for the wedding day. All of the rest is taken care of. "We do it all," says Tony, explaining that they focus on wedding guests rather than the bride and groom. "When the guests have a good time, the bride and groom have a good feeling." To be sure, the Azevedos miss no detail, setting up all equipment and cleaning up afterwards, and even providing a horse-drawn carriage for each bride and groom. Occasionally, problems arise. "Once—when we were expecting three hundred people—the beer cooler broke down. We had to hustle and find half barrels and chill down the beer," laughs Tony. "Whatever problems that might occur, we take care of it. The guests never know, and most of the time the bride and groom are not aware either."

On the whole, however, it's good fun for this couple, who greatly enjoy meeting new people. But it's part-time fun—one or two days per week from April through October. "Anything more would be exhausting," says Tony. "Being a seasonal operation always gives you something to look forward to, by the fall we are exhausted and look forward to closing the season, and in the spring we are excited and ready to open again," says Carol.

In 2005, Tony wanted to expand the museum. The railroad had also had a tremendous impact on farming. Before the railroad, farmers sold their produce within a five- or ten-mile radius. After the railroad arrived, farmers could ship their produce across the country. So Tony decided to add a train exhibit to the museum. The new exhibit, called The History Train, depicts trains from 1880 to 1930, starting with the oldest Union Pacific locomotive still in existence and followed by four other historical cars, one of which is used for dining.

Throughout all of their endeavors, Tony and Carol have continued to operate their dairy. They now farm organically and were one of the first organic dairies in the San Joaquin Valley. They host organic dairy tours for the Ecological Farming Association and reach out internationally as well. When invited to Ethiopia to help farmers suffering in a third year of drought, the Azevedos provided concrete help. "When we came back, we focused on getting thirty-eight thousand pounds of whole powered milk distributed there for immediate aid."

What advice do these busy entrepreneurs have for others entering an agritourism business? "Do what you like doing," says Tony. "And be very aware of your neighbors; farmers need a little time to adjust to a business like ours . . . Keep it educational and agricultural related." Visit their Web site at http://www.thedoublet.com.

Request information about land use and conditions. You'll need

- the zoning map. Establish the exact zone in which your property lies. If you need resource inventory maps to help you identify special features that could impact your plans, ask the county planning department staff members. They usually can direct you to these maps.

- the text listing all allowable uses in your zone and all conditions that affect those uses. Watch for references to other codes or sections and obtain copies of those materials too.

- local, state, and federal requirements and guidelines specific to your activity.

- explanations of any regulation or condition you don't understand.

- an application form for permitted-by-right activities, though it will probably be quite simple.

If, however, your operation requires a use permit, you must learn all you can about the use permit application and approval process.

- Ask for an application for your proposed use and a fee schedule.

- Ask about documentation that must be filed with the application—maps, sketches, letters from the fire district or water district, and so forth.

- Ask about filing deadlines and application approval procedures.

- Ask whether anyone else in the county is doing the same thing as you—and when and how they received permission.

- Ask who makes the decision on your application; counties differ.

Resources for Permits and Licenses

CalGOLD: California Government: On-Line to Desktops, http://www.calgold.ca.gov. Provides California businesses owners with information on permits, licenses, and other requirements of all levels of government agencies.

The California Planners' Book of Lists, http://www. calpin.ca.gov/archives/default.asp. Contains contact information for California city and county planning agencies and is updated annually.

Most county Web sites have downloadable zoning codes, procedures, permit application forms and instructions, and fee schedules. Tour the Web sites of different departments and the services they provide. Start with the departments of planning, building, environmental health, and public works.

- Ask whether any decisions are made "over the counter."
- Ask whether any hearings are needed. Who holds the hearings? Who gets notices of the hearings and the application?

Prepare a Project Description

Before you sit down with county staff to discuss your project in detail, prepare a brief project description and obtain a plot map for your property. Gather additional information, including

- the anticipated number of visitors and months of operation
- the existing and proposed water and sewage or septic systems
- the surrounding land uses
- road access to the property
- timing or schedule issues
- status of Williamson Act contracts

Meet with Agency Staff

It is critical that you meet with representatives of every agency or department that might impact your enterprise. You can either call on each agency individually or visit with representatives of all relevant agencies in a roundtable or predevelopment meeting. Whatever your method, make sure that you and your county share the same understanding about codes and requirements. Listen carefully to objections and concerns. It is easier and cheaper to change your agritourism plans when

you begin the process than it is to fight through appeals after you've spent time and money.

Submit a Complete Application

After you have researched the regulations likely to affect your enterprise, meet with key players to collect the required applications. Assemble and submit your use permit application package to the county planning department. A completed application package usually contains

- a written description of the type and nature of the existing and proposed uses
- multiple copies of the following items drawn to scale:
  - o site plan showing
    - ✿ name, address, and phone number of the owner of record, the applicant, and the engineer and architect if applicable
    - ✿ a north arrow
    - ✿ the date (revised copies should be clearly identified)
    - ✿ the property dimensions
    - ✿ all dimensions of existing and proposed buildings and of additions
    - ✿ the distance of proposed structures and additions to adjacent property lines
    - ✿ parking locations
    - ✿ the existing and proposed topography
    - ✿ inundated areas, streams, culverts, drainage, and swales
  - o building elevations
  - o floor plans
  - o location map showing the subject property to the nearest street

Develop Positive Relationships

Develop positive relationships with the staff of your county planning and environmental health departments. Find out who you work best with and return to that person. In large counties, you'll likely get the person who works the planning counter that day. Clearly explain your plans to them so they understand your intentions. Bear in mind the following:

- Make allies, not adversaries, of your local county officials.
- Ask questions and seek their advice.
- Take time to educate. Remember that few people understand how agriculture actually works.
- Know that regulations are not set in stone; there is often leeway in their interpretation. Some regulations might be out of date or unnecessary.

Putting the Pieces Together: The Use Permit Approval Process

Although each county has its own requirements and procedures, certain steps in the permit process are common to most counties. Because each county has its own permit requirements and fees, it would be difficult if not impossible to list the differences in this publication.

Step One: Exploring the Permitting Process. After you've determined that the county allows your planned activity and that you need a use permit, your next step is to explore the permitting process ahead of you. Meet informally with the county planning department to discuss your plans.

Step Two: Submitting the Application and Forms. After meeting with the planning department and other agencies, use what you learned to complete the required forms. You must submit your use permit application and other pertinent documents to the planning department along with an application fee.

Step Three: Making Sure It's Complete. The planner assigned to your project will make sure your application is complete. California state law requires that the planner complete this step within thirty days of the proposal's submission. The planner will send you an application notice revealing whether your application is complete and, if it is incomplete, what items must be submitted before processing can begin.

California Counties Adapt Permitting and Regulations for Agritourism

California's fifty-eight counties bear the primary responsibility for permitting and regulating agritourism operations on agricultural land within their boundaries. The counties often struggle with creating allowances and ease of permitting for agritourism businesses while ensuring that agritourism is a supplemental (rather than primary) activity on a commercial farm or ranch. Regulations must also ensure that agricultural production and local residents are not adversely affected by tourism. Some counties have recently changed their general plans, zoning ordinances, and staffing assignments to encourage agritourism and have created guides to agritourism permitting.

The Lake County general plan includes Goal AR-3, "To provide opportunities for agritourism that are beneficial to the county and its agricultural industry and are compatible with the long-term viability of agriculture." The county-wide general plan in Calaveras County (Foothill and Mountain region) specifically allows, by right, on-site sales and tasting, and directs that the definition of agricultural operations allowed should be broadly construed. Solano County (Central Valley region) has designated new zoning that encourages agritourism in Suisun Valley, one of ten county regions defined in its general plan.

Mariposa, Placer, and El Dorado counties (Foothill and Mountain region) have involved farmers and ranchers on advisory committees that created ordinances to streamline permitting for agritourism operations while limiting the extent of allowed activities in proportion to the size of the primary agricultural operation.

Potential agritourism operators often complain about the lack of coordinated information from different county regulatory departments. To address this problem, Marin County (North Coast region) contracts with UC Cooperative Extension (UCCE) for an "agricultural ombudsman" to assist applicants with agriculture-related permitting. Marin County UCCE and Placer County staff created plain-language guides for farmstay operations. Yolo County has created an agricultural permit manual that describes all the permits that may be needed for various types of agritourism operations. More coordination among county departments and between counties would ease the regulatory burden on agritourism operators.

Source: Leff 2011.

Advice from County Staff

From county staff comes the following advice to new agri-tourism or nature tourism operators:

- Prepare your business plan or proposal before you meet with county staff. "Make sure you have a clear business plan that describes what you want to do," says an county environmental health specialist. "A regulation is frequently dependent on what the project is—not on the particular industry. It's very difficult to tell an applicant how a regulation applies unless we know what they plan on doing."

- Talk to staff from all agencies that might impact your plans. "We have a process here that we call a predevelopment meeting," notes a county planner. "We invite the applicant to come in and sit down with all of the county departments that would be affected by this proposal or that might have some effect on it. We try to brainstorm the potential concerns or problems and how to solve them—so that when the applicant is ready to actually start or come through the regulatory [permit] process, they've thought through all of these things. We bring the health department in for sewage disposal or water supply, the planning department in for land use conflicts, maybe bring in the public works engineers for any potential access, flooding, and erosion, and possibly the local resource conservation district to talk about erosion and soil loss. If we get a big group of folks who are fairly knowledgeable, then the applicant can ask questions. Or they can be told, 'If you do this, it would cause you this kind of expense. But if you change it just a little bit, lower your sights, and go this way, maybe it only requires this kind of a thing.'"

Step Four: Processing the Application. The planner assigned to process your application will send it along with your plans and other material to county departments and agencies. They review, comment, and provide recommendations. If additional information is needed to fully assess your proposal's impact and conditions, the planner will request it from you.

Step Five: Notifying the Public. Once it is determined that your application is complete, the county planning department will send a notice to landowners near your farm or ranch. This notice will state the date, time, and place of a public hearing.

Step Six: Determining the Action. At the public hearing, your application can be approved, approved with conditions, or denied. The zoning administrator will consider public testimony and any information or comments from relevant departments, agencies, and design review boards.

Step Seven: Appealing the Action. The action of the zoning administrator is final unless appealed. If you appeal the action, you must file your appeal with the planning department within ten calendar days from the date of the action. Check with your county's zoning counter personnel to learn about fees and the appeal process.

How Long Does It Take?

How long it takes to obtain all of the necessary permits varies with the project and agencies involved. A simple project might receive its use permit within a month. Larger projects often take many months.

Surviving Inspections: After the Use Permit Is Issued

Anticipate regular inspections for health or environmental regulations. Get to know these inspectors. A positive relationship makes all the difference in how they perceive your operation.

Potential Permit Issues

When asked what issues arise in the permitting process, county planners gave these examples:

- "California Fish and Game comes back and says, 'We suspect there are rare plants on the site where they are building this.' So we need to hire a botanist to go out and review the area for rare plants."

- "Somebody wants to put a B&B on their farm. Right next door, the neighbor sprays for grapes and says, 'I don't want tourists next to my grape land. They'll be complaining about my spraying.' So, the agricultural commissioner might write back and say, 'Yes, we think it's a bad idea to put it so close to the neighbor's operation.'"

- "Neighbors and maybe the department of public works voice concern about the traffic on a certain road. If you're going to be inviting tourists to come up the road, all of the sudden you've got increased traffic—and maybe the intersection is dangerous and needs to be modified."

Permits, Licenses, and Your Business Plan

Estimate Permit and Licensing Fees

When you finalize your business plan, you must include the costs of permits and licenses. Most counties have fees associated with the permit application process. Licenses, too, often require yearly renewal fees.

Frequently, a fee accompanies your permit application and another fee is required for your public hearing. Fees usually change annually, often after July 1 (the beginning of the fiscal year for most counties).

In addition to the North American Farmers Direct Marketing Association, there are statewide support and networking associations there to help you. These associations can be a great resource for advocating for zoning and permitting changes to county regulatory codes. Here are a few examples.

- Central Coast Agritourism Council, http://www.agadventures.org/

- Hawaii Agritourism Association, http://www.hiagtourism.org/

- North Carolina Agritourism Networking Association, http://ncana.blogspot.com/

- Vermont Farms! Association, http://www.vtfarms.org/

You can find a complete list complied in 2007 by Lisa Chase from University of Vermont Extension and her colleagues from Utah State and University of Michigan at the Iowa State Agricultural Marketing Resource Center (AgMRC), a great online resource at http://www.agmrc.org.

Estimate the Costs of Compliance

You are responsible for all costs. If you modify your facilities to meet use permit conditions or construct a new barn in compliance with building codes, you must pay the costs. Therefore, it makes sense to identify regulatory requirements early on so you can anticipate expenses and build them into your business plan. You can use the worksheet in figure 4.1.

Figure 4.1

Regulatory Agency Worksheet

Jurisdiction	Agency	Permit or Fee	Start-up Cost	Annual Cost
County	Agricultural Commissioner	Scale certification		
		Organic inspection		
		Labeling inspection		
	Treasurer/Tax Collector	Business tax and license		
		Transient occupancy tax registration		
	Building Department	Building permit		
	Planning Department	Use permit		
	Environmental Health Department	Food safety permit		
		Food processing permit		
		Increased or new septic permit		
		Increased or new well permit		
	Public Works	Entry onto county roads permit		
		Signs permit		
	Recorder	Fictitious name filing		
	Fire Agency	Safety and occupancy inspections		
State	California Franchise Tax Board	Income taxes		
	California State Board of Equalization	Sales and use tax registration		
	California Department of Alcoholic Beverage Control	Liquor license		
	California Department of Fish and Game	Fishing, hunting permits		
	California State Department of Industrial Relations	California Occupational Safety and Health Administration: Employee safety plan		
	California Department of Transportation	Entry onto state highway		
		Signage license and fees		
Federal	Bureau of Land Management	Special-use permit		
	U.S. Forest Service	Special-use permit		

Points to Remember

- Numerous regulations—many of them complex—face landowners interested in establishing an agritourism or nature tourism enterprise.

- Regulations are part of doing business, and your compliance with them helps protect you as well as your customers from potential liabilities.

- You can best address the regulatory bureaucracy by taking it one step at a time.

- Agency Web sites and staff can answer questions, provide information, and help you meet requirements.

- A good working relationship with all agency staff is vital, both during the permit application process and during later inspections.

- The time required for the permit approval process varies with each operation; therefore, allow for a lengthy procedure.

- Regulations are important to the development and cost estimates of a business plan, so identify them early in your planning.

The Reality of Cost

"These farm-related businesses are not huge projects, so even the cost of a $10,000 or $20,000 roadway improvement is significant to these folks," says a county planner. "That's where we've had some difficulty—because, frankly, we don't control the cost of asphalt. If we absolutely need that widening of the road, then that requires some grading and some asphalt. If we feel from a safety standpoint that it's necessary, we've got to find some way to get it done.

"The application fee just covers the cost of public notice, the time the staff puts in, and so forth," notes another planner. But, say, the archaeological commission comes back and says, 'We think there's a Native American site on the property that needs to be looked at.' The applicant has to hire an archaeologist, pay them a couple hundred dollars to go out, look, and file a report. Those types of studies might cost money."

"The whole permitting process—with the regulations—probably increased the cost of our new building by 50 percent," observes a farmstay operator.

Chapter 5

Developing Your Risk Management Plan

Chapter Goals

The goals of this chapter are to help landowners

- understand the legal responsibility of agritourism and nature tourism operators
- realize the importance of a risk management plan to business success
- develop a risk management plan to limit landowner risk and liability
- identify strategies to reduce farm-safety risks and financial risks to visitors, employees, and employers
- identify hazards on the farm or ranch
- recognize the need for liability insurance and which type of coverage to purchase
- draft a liability waiver
- understand labor issues
- employ good employee-management practices
- address the special needs of children, the elderly, and disabled people

About Landowner Liability

Liability—legal responsibility—should be of concern to anyone contemplating a tourism enterprise. You'll need to consider two main types: premise liability, in which someone is injured or harmed on your property, and product liability, in which someone is sickened or harmed by a food or product you supplied. According to California law, the person who owns or leases property is the person who is in control of that property and is therefore the person legally responsible, or liable, for the well-being of visitors.

Tractors are not passenger vehicles. They are built for one person to control and perform specific tasks. A person other than the operator who hops a ride aboard a tractor or trailing equipment can fall off or get caught in the machinery, becoming seriously injured or even killed. A "no riders" policy for tractors and other farm equipment is especially important for farms that host visitors, offer hay rides, or demonstrate equipment.

Operators often think they can stop the tractor in an accident, especially if the tractor is moving very slowly or no difficult tasks are being performed. The most common comment from people involved in tractor runovers is how quickly they happen.

What Is a Risk Management Plan?

Risk management is a way to reduce legal liability through planning and business management. In your own agritourism or nature tourism enterprise you will face two kinds of risks: farm safety risks and financial risks. Farm safety risks involve your physical operations. Financial risks involve employee issues such as loss of a major customer, injury of a guest, or closure of a farmers' market where you sell your products.

Craig Raysor, an agricultural and food attorney with the firm of Gillon and Associates, PLLC, in Memphis, Tennessee, talks about the legal considerations you need to take into account before you begin inviting people into the barn or out into the field.

"First, you are held to a higher duty of care since you invited these people onto your land, whether for payment or free of charge, than if you just allowed a friend on the land to kick back with you or if they trespassed. The duty of care in the invitee situation, which you have in agritourism, is for you to use ordinary care to keep the premises reasonably safe for the benefit of the invitee. This means you are held to the same liability that the Wal-Mart in your town is held to regarding customer's expectations. Therefore, it is important that you post where the invited guests are allowed to be on the farm as a protection against further liability. This also means that you need to have safe equipment, and properly trained personnel operating that equipment. Do not let your nine-year-old give a ride to the guests around the farm on a barely running tractor.

"Please research your state statutes or, even more advisable, hire a specially trained attorney, to look over the state statute to see if you fit under a recreational use statute if you do not charge for the person to be on the land. You may be free of all liability, except egregious or intentional acts, if your operation fits under a recreational use statute.

"Second, see if your current insurance policies cover such activity. You may find many do not cover agritourism activities, because of the higher care and the higher likelihood of injury. You may have to get additional coverage or a separate policy altogether to protect your farm. You may have to do some additional digging in this arena, as many of your local carriers may not be able to offer such coverage.

"Third, you may want to create a separate business formation for tax purposes and liability protection for your agritourism activity. This will allow only assets attributed to that company to be at risk in case of a lawsuit. There are a variety of different formation options that can be dictated by state law. There may even be encouragement within your state to form agritourism cooperatives as it has been in many southern states.

"Fourth, check out your state's ag department if they are getting behind the value-added product of agritourism. There are grants and matching funds programs out there within the states and federal government. In addition, it can be a wonderful marketing tool for your farm to utilize in addition to your individual marketing. It has even become a tab on the general Tennessee tourism site with links to individual farms throughout the state."

These are a few ideas and suggestions to get you going, but remember to have fun with the new venture as well and use it as a time to let your farm and its products shine.

Source: Smart 2009.

Although you cannot eliminate these risks, you can reduce them as well as your liability:

- Plan ahead; meet with appropriate professionals and keep accurate records.
- Avoid certain activities.
- Exclude visitors from potentially risky places and use signs.
- Make your operation as safe as possible.
- Issue liability waivers.
- Select the appropriate legal structure for your business.
- Make sure your products are labeled properly.
- Learn and follow good management rules about hiring, training, and developing employees.
- Buy adequate insurance.

Together, these strategies form a risk management plan. Creating such a plan requires that you think and act proactively, and that you identify your assets and reduce their associated hazards. Should an accident occur, your risk management plan can help to legally protect you. In 2010, nineteen states in the United States had enacted statutes that address agritourism. These statutes vary from liability protections for agritourism operators to tax credits to zoning requirements. The Indiana state senate passed a bill in 2011 limiting agritourism liability. The bill assumes participants in agritourism events, from apple picking to hay rides, understand the risks associated with the activity they're engaged in. The bill's sponsor, Bedford senator Brent Steele, said the liability does not include everything. "It does not relieve the farmer from liability for those known or dangerous conditions that are not in an ordinary course of the farming operation," said Steele (Smith 2011). See the National Ag Law Center Web site, http://www.nationalaglawcenter.org/, for a wealth of information about agritourism and risk management.

Since many agritourism visitors are children, consideration should be given to eliminating attractive nuisances and taking extra steps to ensure health and safety. The Childhood Agricultural Safety Network is working to spread the message "Keep Kids Away From Tractors," and the UC Small Farm Program Web site has a number of factsheets about safety and risk management for agritourism.

The following pages outline strategies to ensure the safety and well-being of your visitors and employees.

Risk Management Plan, Part I: Farm Safety

To decrease your operation's physical hazards, start by practicing the farm safety steps that follow.

Farm-Safety Strategy No. 1: Reevaluate Your Venture

The goal of an agritourism or nature tourism endeavor is to attract visitors. Tourism enterprises can be risky businesses—not only for visitors, but for staff, operators, and animals, too. Because of this risk, carefully contemplate the establishment of a tourism enterprise.

Farm-Safety Strategy No. 2: Reconsider Your Activities

What about visitor activities? Reexamine them. Identify activities that are potentially hazardous and avoid them. Look at your other activities and create a plan to ensure visitor safety.

If you're planning to sell processed foods, make sure you comply with state and local rules for licensing food-processing facilities.

If you're planning to offer hay rides, make sure you identify safety precautions and insist that visitors follow them. For instance, you might require visitors to stay seated and keep their arms and legs far from wheels. You might also limit the number of riders, drive your tractor as slowly as you can jog, and regularly check your route for potholes, irrigation flooding, and other hazards.

Farm-Safety Strategy No. 3: Make Your Place Safe

Look around. That pothole you dodge every day in your driveway could damage a visitor's car. That gate that you jiggle shut could hurt a child climbing on it. A rusty nail in a fence could scrape a visitor. Your family dog could bite a stranger. Identify and reduce the dangers facing visitors unfamiliar with your farm or ranch. Make your property safe!

Reduce physical hazards
Post boundaries. Post property boundaries and "No Trespassing" signs. Block access to areas

closed to visitors (i.e., equipment shop, pond, etc.). Provide every guest a map showing visitor-designated areas.

Identify and enforce visitor areas and rules. Take a good look at your operation: how it's laid out, how it's run, and how buildings and facilities are designed. Armed with this information, define visitor areas, visitor activities, and visitor supervision. Establish safety rules and implement them.

Post safety rules. Post and enforce your safety rules. Signs might include "No Smoking," "No Alcohol," and "Close All Gates."

Clean up visitor areas. Fill that pothole; fix that gate; pull those rusty nails; build a dog run! Make sure your property is safe. Repair damaged objects and demolish unused outbuildings. Install equipment and facilities to prevent personal injury or property damage—mount railings on your stairs and place antislip pads on your steps. Maintain those railings, stairs, and steps! Also regularly inspect furnaces and heaters and frequently check and repair fences. Good lighting is an important safety feature: install yard lights and other lights where needed and furnish flashlights where appropriate.

Explain hazards to your guests. Waste no time in telling your guests about the risks associated with your farm or ranch. Explain that you run a working operation and hazards like biting insects, poisonous snakes, inclement weather, farm odors, unpredictable farm animals, and rough terrain exist. Describe these hazards in detail and explain that visitors must accept the risks and exercise reasonable caution. Hand out a written warning describing all known risks. Post warning signs. Lead a health and safety tour of your premises. You should also require that visitors wear appropriate clothing, perhaps long pants and closed-toed shoes rather than sandals for certain activities.

Conduct safety-education programs. Provide safety-education classes for employees on a regular basis.

Guard against fire. Maintain defensible space around buildings to reduce fire danger, conduct fire-prevention inspections, and install fire extinguishers where needed and where required by law. You must also install an approved and properly functioning smoke detector in every California guest room, according to state law.

Establish an emergency plan. Establish an emergency plan for fires, floods, earthquakes, and other natural disasters. Keep well-stocked first-aid kits handy and provide first aid and CPR training to all employees. At the same time, post emergency telephone numbers where everyone can see them. It is equally important that you make sure local emergency crews know how to get to your place.

Block off bodies of water. Block off ponds, lagoons, rivers, and streams from the public, using fencing, ropes, or even cones. Because children gravitate toward water, caution their parents to supervise them, even if you have blocked access to the water.

Provide adequate parking. Most counties require that cars be parked off paved roads. Make sure you have enough space for the number of vehicles you expect.

Plan convenient transportation. If buses park away from your farm, provide a drop-off and loading location.

Be secure. Depending on the event you're holding, hire additional people to make sure your guests are safe and your farm is secure.

Provide sanitary restrooms. Do you have public restrooms that are clean, well stocked, and in good working condition? If you plan to have numerous visitors and employees, you might rent portable toilets.

Provide hand-washing facilities. Provide hand-washing facilities within and next to areas of animal contact. Hand-washing facilities include running water, soap, and disposable towels. Hand sanitizer gel packs are acceptable, but their effectiveness in animal settings is uncertain. Baby wipes are not an acceptable alternative. Explain to guests that they must wash their hands after touching animals or visiting animal exhibits and dwellings. Post signs saying, "Wash your hands after handling animals."

Bar visitors from animals not available for viewing. For the health and safety of both visitors and livestock, prevent their interaction. Place animals that are not available for viewing away from visitor areas.

Prevent hand-to-mouth activities where animals are located. In areas of animal contact, do not permit hand-to-mouth activities such as eating, drinking, smoking, and carrying toys and pacifiers.

Safely store pesticides and other poisons. Store pesticides, herbicides, and other toxins in a locked building and discard hazardous waste according to environmental regulations.

Lock your shops. Shops and repair facilities are hazardous places. Keep them off-limits—shut doors, rope off entrances, and hang "Do Not Enter" signs. You should check these areas often.

Keep implements and equipment away from visitors. Agricultural equipment and implements fascinate some people. Make sure that visitors respect them: don't allow children to climb on your property and never allow visitors to drive implements. Park all farm vehicles away from visitor areas. Hide ladders to eliminate the temptation to climb.

Be vigilant. Be on constant lookout for hazards and promptly address them.

Reduce risks from and to animals

Identify how you can minimize disease transmission from animals to people and from people to animals. Pathogens can enter your farm and ranch on clothing and personal items. Practice good sanitation for the health and safety of your animals as well as your guests. Animals are unpredictable and often behave differently around crowds of people. To limit the liability associated with animals, supervise all interactions they have with guests. Confine your "viewing animals," allowing visitors only limited and controlled access. Make sure you choose your healthiest and friendliest animals for public interaction. Always consider animal well-being to be paramount to your operation.

Make sure you vaccinate viewing animals against rabies; if animals are too young to vaccinate, exclude them from your exhibit. Exclude ill animals as well, especially those with diarrhea or signs of encephalitis such as stumbling, lack of coordination, or paralysis. You must immediately report these conditions to the local health department, as well as visitors being bitten or scratched. You should make an effort to control odors, manure, flies, and other pests and provide good ventilation, particularly in the visitor area.

Farm animal display resources

The following list from the North America Farmers Direct Marketing Association provides ideas to reduce risk and liability for specific animals.

Cats and dogs. Make sure that only friendly, social dogs are near the public. Warn visitors about puppies' sharp teeth; even friendly puppies can do damage! Warn visitors about cats' sharp claws and teeth as well.

Birds. Reduce your birds' stress by limiting their interaction with people—in other words, rotate your viewing birds. Safeguard your guests, too. Although you might allow visitors to feed chickens, ducks, and other poultry, keep in mind that geese can be aggressive.

Small livestock. Consider using goats and sheep as petting animals rather than horses and ponies. Goats and sheep are smaller, lighter (and less injurious) if they step on a guest, and they lack top front teeth with which to bite. In general, they and other small livestock are safer than larger animals.

Horses and ponies. Horses and ponies bite, kick, step on toes, and buck. Warn your guests! If you plan to offer horseback riding in any form, see your insurance agent about liability protection.

Cattle and calves. Cattle kick, too! Control calves and cows that will be handled. Do not allow hand-milking.

Alcoholic beverages

A landowner selling alcohol must carry liquor legal liability insurance. See chapter 4, "Understanding and Navigating Regulations."

Transportation

A landowner providing transportation in a personal vehicle requires a commercial automobile policy. Regular coverage will not cover the vehicle's commercial use.

Vicarious liability

A landowner is responsible not only for his or her own actions but also for those of people acting on the landowner's behalf. In other words, you are responsible for the actions of your employees and of independent contractors such as pack or hunting guides. You should require independent contractors to maintain their own liability insurance if they are able to do so and have them submit written proof of this to you, as well as their statement that they are responsible for their workers' compensation insurance. This is not always possible; in these situations, you need to have your liability policy endorsed to cover your hunting guides as additional insureds.

Off-farm liability

Liability protection different from that already described is needed for farm-sponsored activities that cause off-farm harm, hazard, or injury. These activities might include visitors trespassing onto adjoining property, chemicals drifting or fire spreading onto your neighbor's property, or water seeping onto adjacent land. Some tourism activities might cause "nuisance" to an off-farm party as well. Again, make sure you discuss your potential liability with both your lawyer and insurance agent.

Address other liability issues

Paid recreational events and activities often involve different kinds of liability and, therefore, different kinds of liability protection; check with your agent.

Farm-Safety Strategy No. 4: Issue Liability Waivers

Although it does not absolve you of responsibility for your guests' health and safety, a liability waiver is a legal document that is valid in court. It is also a terrific educational tool, highlighting the potential risks facing your visitor. Liability waivers are especially important for visitors participating in high-risk activities such as horseback riding. Each state views liability waivers differently, so check with your lawyer or insurance agent.

Drafting a liability waiver

The following suggestions can help you draft a liability waiver. Your lawyer can provide valuable assistance as well.

Type of agreement. Identify the type of property use. Define the terms under which a person or group may enter your property.

Names of the parties. Leave space on the form for each visitor to fill in his or her name and address.

Description of tract. Clearly describe the size and boundaries of the land that the liability waiver covers. Include areas that are off-limits as well as safety zones around houses, barns, and pastures.

Statement of purpose. Clearly and specifically describe the uses permitted under the agreement.

Facilities and services provided. Describe food, lodging, guides, transportation, and more.

Activities. Identify and describe the activities you offer.

State law compliance. Note that the visitor agrees to comply with all applicable state laws while on the landowner's property.

Registration. Require that all visitors check in and check out at one specified location.

Limitations. Explain the right of access to adjoining land.

Indemnity by user. Specify that the user agrees to "indemnify and hold harmless the landowner from any claims made by the user or third parties arising from the use of the land or activities." This means that your guest agrees not to sue or otherwise attempt to make claims against you as the owner regarding his or her use of your land and activities on your property.

Waiver of liability. For additional protection, incorporate firearm-safety principles, require successful completion of hunter-safety or horseback-riding basics, and prohibit alcohol consumption.

Acknowledgment and assumption of risk. Describe your property, recording, for example, "wildland with barbed-wire fences, logs, poison oak, wild animals, and uneven terrain." Also describe your activities, noting that the guest accepts risks associated with the activities and property.

Insurance. Specify that the visitors carry liability insurance and, in their policy, that they name the landowner as an insured.

Conditions of cancellation, renegotiation, or renewal. List necessary conditions of cancellation, renegotiation, or renewal in case either party terminates, renegotiates, or renews an agreement for various reasons.

Mediating differences. Reduce the risk of disagreements resulting in litigation. Specify that problems—injuries included—arising from the waiver of liability and the use of property will be addressed in mediation before either party resorts to legal action.

Payment. Specify the payment amount, payment method—where, when, how—and provisions for failure to pay.

Damage deposit. Specify a deposit for damages incurred by the visitor and not repaired by the visitor, noting that the deposit is returned if damages do not occur.

Other concerns. You might want to address additional concerns. For instance, you might cover trespass enforcement, gate and fence care, in-kind services that the guest provides, limitations of the number of campers in a particular area, campfire use, garbage removal, and sanitation requirements.

Signatures. Complete the agreement with the printed name and address of your visitors, followed by their signatures and the date of their signings.

Incident reports. Keep a log of your safety inspections and sign off on it. Keep an incident report that documents any incident that occurs with your employees or a visitor. Collect names, contact information, and a statement from the visitor if necessary.

Figure 5.1

Permission to Enter and Use Property

_____

(guest)

is hereby given permission to enter property known as

_____

for the following purposes:

_____

_____

on the following dates: _____

This permission is limited strictly to the above description. It is subject to strict compliance with the rules and regulations copied and attached and with other limitations or restrictions that occasionally may be issued either orally or in writing.

Release, Waiver, and Indemnity

I have read the above permission and the accompanying rules and warnings. I understand that participating in

_____

(name of activity)

results in certain risks, regardless of all feasible safety measures taken. I am aware of the risks involved. I will follow all rules presented to me. I will conduct myself as a prudent person with regard for the safety of myself and others and of the property of others.

To the extent that proposed activities involve "equine activity" as defined in state law, I hereby waive the right to bring action against the equine professional or equine activity sponsor for any injury or death arising out of riding, training, driving, grooming, or as a passenger upon the equine.

I assume the risk of any responsibility for injury, loss, or damage to person or property resulting from my participating in activities on the premises. I will not hold the landowners, possessors, or occupiers liable.

I agree to indemnify and hold the landowners, possessors, and occupiers harmless from any claims or damages resulting from my actions that may affect the person or property of the landowners, possessors, or occupiers of the premises or any other person.

Signed _____ Date _____

A sample waiver

Figure 5.1 is a sample liability waiver with elements that might be useful; other examples can be found on the Internet. Draw up your own waiver and photocopy it for your visitors.

Farm-Safety Strategy No. 5: Choose the Appropriate Legal Structure

Like it or not, Americans are litigation happy. You can take every precaution imaginable and still find yourself in a lawsuit. So, structure your business in the way that best protects you.

For example, a customer might step in a gopher hole and break a leg, a turkey might peck a child, a cow might kick a guest milking her, a U-pick customer might amble into poison oak, or an employee might improperly load a pumpkin and damage a visitor's car. Any one of these situations, cited in the North American Farmers' Direct Marketing Association Web site, http://www.nafdma.com/, can result in legal action.

Before you host a single visitor or plan even one event, meet with your management team (see chapter 3, "Creating Your Business Plan").

Your lawyer, who is familiar with business law, and your insurance agent are of particular importance to your risk management strategy.

With their assistance, you can decide on a legal structure for your business to reduce liability. The structure you choose will influence many components of your enterprise, including management, taxation, and estate planning. Will you decide on sole proprietorship, general partnership, limited partnership, corporation, limited liability company, or limited liability partnership? All of these options are described in chapter 3, "Creating Your Business Plan."

Farm-Safety Strategy No. 6: Buy Insurance

Insurance needs vary by operation. Knowing what your risks are helps you evaluate what type and how much insurance you need. Chapter 7, "Resources for Success," lists a host of resources, including Neil Hamilton's book *The Legal Guide for Direct Farm Marketing*.

Consider liability insurance, which shifts the risk from you to your insurance company. If you've taken steps to prevent an accident, yet an accident still occurs, your insurance policy should adequately cover you. It should protect your assets and—if you face a lawsuit—pay for added costs. As Eric Gibson explains in *Sell What You Sow! The Grower's Guide to Successful Produce Marketing*, such costs might be for helping the victim, the investigation, the defense or settlement, and the court bonds or interest on judgment delayed by appeals. Without liability insurance, you risk everything you own.

It is critical that you talk candidly with your lawyer and insurance agent to determine whether your current coverage is adequate. You want the right kinds of insurance and the right amount of coverage. Most likely, you'll need additional coverage, which means having to add a rider to your existing policy or buy a new policy with a different company.

Be aware that most farm and ranch policies discourage agricultural and nature tourism activities because of work site dangers. In general, insurance companies consider one occasional guest acceptable but a stream of paying visitors objectionable. As a result, they either add exclusions to their policies or provide special, more expensive policies.

Purchasing coverage

When asked about liability insurance and other risk management practices, 87 percent of the California survey respondents reported having liability insurance in 2008, and 90 percent of the insured were covered for $1 million or more (Rilla et al. 2011). Several people commented about the cost of liability insurance and concerns about being sued.

When you do purchase insurance, choose a package that protects your personal assets. The cost varies with the company, activities, and risks facing your visitors. It also reflects the share of your entire operation that your new enterprise encompasses.

Make sure you employ a quality insurance company and that you understand your coverage. And—like all other aspects of your enterprise—make sure you review your coverage at least once a year and get a competitive quote every three to four years.

Start with your current insurance provider, then contact other companies and agritourism and nature tourist operators, asking plenty of questions. The form in figure 5.2 can help you track their answers.

Where to start?

The USDA Agricultural Marketing Resource Center Web site, http://www.agmrc.org/, includes information about agritourism insurance and liability issues, providers, and sample indemnification agreements. The North American Farmers Direct Marketing Association, http://www.nafdma.com/, maintains an insurance company referral list. Traditional farm commercial liability programs usually exclude coverage for educational and recreational activities undertaken for compensation, but the new Agribusiness Commercial General Liability (AgGL) program covers B&Bs, mazes, tours, hunting and fishing, farm parks, and other activities offered by agritourism operators.

In 2004, the American Farm Bureau Federation recognized agritourism as an on-farm activity, which may help operators who are using their state farm bureau insurance to cover new activities. A number of states have legislation that limits liability to agritourism operators. You can read a short overview of some of these laws at the Arkansas Agritourism Web site, http://www.aragriculture.org/aai/initiative/legislation.asp.

The sources below can get you on your way to insuring your tourism operation. Although the list is neither complete nor an endorsement, it does name companies in the western United States that cover at least some agritourism and nature tourism activities. Please be aware that the insurance industry is fluid and that some of these companies may no longer provide coverage.

- California Farm Bureau Federation members can work with some providers available from local agents representing Allied Insurance (http://www.alliedinsurance.com/) and Nationwide Insurance (http://www.nationwide.com/) or by calling toll-free 877-OnYourSide.

- Allied Insurance and Nationwide Insurance provide coverage for California Farm Bureau Federation members, covering farm or ranch property and liability.

- Gillingham and Associates, Inc., http://www.phly.com/products/OutdoorProducts.aspx, provides coverage for outfitters and guides, horses and related activities, guest ranches, fishing, hunting, trap and skeet shooting, field

Figure 5.2

Buying Liability Insurance

Call several companies to compare rates and coverages.

Insurance Company:

	Yes	No
Is there a deductible?		
If yes, how much?		
Does the insurance apply to		
my premises and operations liability?		
my products and operations liability?		
my contractual liability to others?		
my personal injury liability to others (libel, slander, invasion of privacy)?		
my advertising injury to others?		
my property liability damage to others?		
incidental medical malpractice liability resulting from my helping an injured person?		
non-owned watercraft liability?		
host-liquor liability?		
court costs for defense—above limit or included in liability policy limit?		
Is each of my employees added as an additional insured?		
Is the premium a set fee?		
Is the premium based on a percentage of gross sales or on client days?		
Do I have to belong to an association to purchase insurance?		
If yes, what is the cost of membership?		
Does the insurance agent understand my proposed enterprise?		

Notes

dog training, hunting preserves, shooting clubs, snowmobile tours, and related activities. Call toll-free 800-849-9288.

- Worldwide Outfitter and Guides Association, Inc., 800-321-1493, provides liability insurance to its members as participants in a group master policy. It offers coverage for outfitter and guide services and for closely related activities.

Risk Management Plan, Part II: Employee Well-Being

The second part of your risk management plan is financial well-being specifically related to employee issues and strategies to reduce your employee-related risks and liabilities.

Risk management regarding employees involves good sense and sound information. States such as California have many laws covering personnel issues and contracting, so it's imperative you talk with your lawyer. Note that good management practices are of primary importance in this area. When you learn and follow good hiring and training practices and safeguard your employees' health and safety, you are well on your way to protecting yourself legally. See chapter 6, "Designing Your Marketing Strategy," and chapter 7, "Resources for Success."

Employee Well-Being Strategy No. 1: Acquire Employer Status

Begin contacting the federal and state governments for your employer identification numbers.

Employee Well-Being Strategy No. 2: Follow Good Hiring Practices

Ask yourself the following questions.

- Can I add my new enterprise without overloading my existing staff?
- How will I recruit additional staff?
- What process will I use to screen and hire employees?
- What training will I provide, by whom, and at what cost?
- How will I set salaries and wages?
- What benefits will I provide?

Start by listing the employees your enterprise needs. Note their titles, duties, and skills. Research the pay range offered elsewhere for similar work. Then, if you know a friendly, knowledgeable person who would reflect favorably on your business, ask them if they'd like to work for you. If you offer part-time jobs and seasonal jobs, your community may have responsible residents who would like to "help out."

If you advertise for employees, decide how and where you'll attract them. You must advertise the position as required by law and supply applications and a job description to everyone interested.

How will you screen applicants? Do you need a course in effective interviewing? Interview qualified people and thoroughly check references of interesting candidates. Always abide by equal employment opportunity requirements: see the U.S. Equal Employment Opportunity Commission Web site (http://www.eeoc.gov/) and the California Department of Fair Employment and Housing Web site (http://www.dfeh.ca.gov/DFEH/default/).

Once you've located the ideal candidate and are ready to offer employment, include in the job offer

- a statement of wages or salary
- the number of hours to be worked
- the days and times to be worked
- pay days
- benefits

Ten Rules of Good Business

1. Customers are the most important people in our agritourism and nature tourism enterprise.

2. Customers are not dependent on us; we are dependent on them.

3. Customers are not an interruption in our work; they are the purpose of it.

4. Customers are doing us a favor when they call; we are not doing them a favor by serving them.

5. Customers are part of our business; they are not outsiders.

6. Customers are human beings with feelings and emotions like our own; they are not cold statistics.

7. Customers are not to be argued with.

8. Customers are bringing us their wants; it is our job to address those wants.

9. Customers are deserving of the most courteous and attentive treatment we can give.

10. Customers are the lifeblood of the agricultural and nature tourism industry.

Risk Management Case Study

Impossible Acres Farm

Just outside of Davis, California, Katie and Clyde Kelly run Impossible Acres Farm—their family farm and agritourism business. In spring, guests come seeking fresh produce. In summer, they pick fruit and berries. And in fall, they harvest pumpkins, ride the hay wagon, and pet the animals. All the while, the Impossible Acres' red barn at the nearby busy county road intersection draws in customers. On that same corner, a fruit stand (with walk-in cooler) serves fresh fruit to visitors wanting to buy rather than pick farm produce.

October is the busiest month for the Kelly family, attracting as many guests as all other months combined. By offering different activities in different seasons, however, the Kellys can support themselves throughout the year. What's more, by providing educational activities, they can encourage the farming industry. Their hands-on learning activities give urban residents insight into the importance of farms and farming.

To first- and second-grade students, Impossible Acres Farm offers activities that Katie and Clyde have designed with a preschool teacher. Children learn where their food comes from, see baby animals, visit the pumpkin patch, ride the hay wagon, and run the hay-bale maze. Soon, the farm will offer opportunities for older students too. The Kellys are planning a program that would, for example, show older children how a tomato grows by having them plant seeds, transplant seedlings, weed, water, and harvest—all in one visit.

Although all activities on the farm are fun and child-friendly, the issue of liability looms. Therefore, Katie and Clyde have taken concrete steps to limit liability, planning and organizing their farm to reduce hazards, particularly for children. For example, there are no pumpkin piles for children to climb and no dangerous animals. The Kellys have also created farm policies and enforce them. That is, they accept school groups by reservation only and require that each visiting group includes one parent supervisor for every four students. When the school group arrives, Katie immediately discusses farm rules with all participants.

Another liability-limiting step the Kellys have taken is to procure liability insurance. Under an umbrella policy with Calfarm insurance, Katie and Clyde have insured Impossible Acres Farm at approximately $2,000 per year—a price Katie considers reasonable. To reduce the insurance cost, they've taken extra precautions. They allow no guests to climb ladders while picking fruit, instead growing dwarf trees to make the job easy. They wrap all wire in visible colored tape. And they post clear and obvious instructions and warnings.

To be sure, visitor safety even impacts the couple's farm practices, such as the choice of organic versus conventional growing. Although the Kellys work actively to conserve resources and care for their land, they recognize the occasional need for conventional practices. "A black widow is a horror scene!" says Katie. "If we find black widows, we've got to get rid of them. If we seem to have an infestation in a certain area, we're going to spray."

Because it's a farm with small animals, tense situations inevitably arise. When they do, it's best to respond with caring and kindness, says Katie. "Be as nice as you can. Generally, guests just want to know that you'll take care of the problem, that you're concerned."

She recalls the woman bitten by a kitten last year. "They were pulling the cat, and the cat was tired and bit them. The lady was afraid that the cat might have a disease. The issue then became not 'Does the cat have a disease?'—because we knew it was fine—but 'How do we deal with her concern?'"

There is so much to learn, Katie and Clyde say. It helps to have a strong working relationship among partners, acknowledging and using each other's strengths and weaknesses. It helps to have a good working model as well. The Kellys lean on Katie's father's U-pick operation at San Luis Obispo, in Central California. His ten-year head start allows them to watch and assess his progress and weigh their options.

Whatever their future path, Katie and Clyde promote their current services. Their marketing pays off. Recently, they posted a Web site, http://www.impossibleacres.com, that has already expanded the distance guests travel to visit. For example, it has attracted a family from Reno, Nevada—three hours away—and more people than ever from the San Francisco Bay Area, over an hour west.

Katie and Clyde use the local media as well. With it, they invite customers to come harvest overripened crops and to attend special events. When they built their new barn, they issued a press release inviting public participation. Their press release launched a relationship with the local press: "Now the reporters know us!" Moreover, their invitation strengthened their bond with the community, increased their farm's exposure, and promoted urban-rural communication. Visit their Web site at http://www.impossibleacres.com.

Desmond Jolly, Director Emeritus, UC Small Farm Center, and Isabella Kentfield, Research Assistant, CalAgVentures

At the time of the job offer, you might also obtain a young employee's proof of age, an immigrant's work visa, a minor's work permit, and, from everyone, their Social Security numbers, I-9 forms, W-4 forms, and a new hire report (DE54) for the California Employment Development Department. You must provide your employees with unemployment insurance, state disability insurance, and workers' compensation, according to law. You are not necessarily required to provide all employees health insurance, however, depending on circumstances such as how many employees. Refer to chapter 7, "Resources for Success," for more information.

Employee Well-Being Strategy No. 3: Orient Your Employees

Orient all employees, new and advancing. Take every employee on a tour of your operation, pointing out how they fit into the "big picture." Regardless of position, every person on your operation is your ambassador and must know and understand your business. Explain to them why you pursued a tourism enterprise, why you consider it important, and what everyone involved must do to prosper. Encourage questions—any time, any day—so that your staff keeps informed and feels like a team.

When new employees arrive, introduce them to your working team. Show them your equipment, point out storage locations, and discuss rules. If you have an employee manual, hand it out, along with a job description, task list, and safety rules. Make sure they understand the importance of customer service. Explain that customers are not interruptions but, rather, the very reason you're in business and able to hire staff.

Employee Well-Being Strategy No. 4: Train Your Employees

Train and retrain your employees, family included! Trained and educated workers are more confident, competent, and self-directed. Teach them about your enterprise and educate them about the tourism industry. Provide videos, publications, and access to outside programs, plus any needed time off for classes.

You must decide who will educate your workers and at what cost. Will you go through a consultant, community college, or other recreational business? Or will you train them yourself? If you train your staff yourself, first learn how. Take a course, watch videos, listen to tapes, read books, and talk to people who know. To help clarify your thinking and generate a training guide, write down the responsibilities of each position and the expected outcomes.

Above all, remember that no one will do the job like you do. Even with all of their training, employees might do the job as well or perhaps even better than you, but rarely will they do it exactly like you do. Consequently, you must manage employees for results, not methods—unless safety, health, or your own reputation is at risk. The questions you should ask are Did the job get done? Was the job done well? Are my customers continually satisfied by my employees' work?

Employee Well-Being Strategy No. 5: Encourage, Empower, Reward!

Encourage, empower, and reward your employees. Treat them with respect and as you want them to treat your customers. Your effort will be reflected in their morale and behavior. It is important you show your workers that they are crucial to your success. Hand them responsibility. Tell them and show them that you value their work.

Thank them for a job well done, perhaps in writing. Share customer comments about their service. Convey your pride and post a noticeable display with their photographs and descriptions of themselves and their interests. (This also furnishes your customers a comfortable sense of familiarity.)

Meet regularly with your staff, such as in a 15-minute meeting at the beginning of each work week. Review the events of the prior week and expected plans for the week ahead. Discuss customer comments, complaints, and suggestions. If the environment of these meetings is comfortable and open, relationships, trust, team spirit, and teamwork will grow.

Employee Well-Being Strategy No. 6: Pay Fair Wages

Pay your employees a fair wage, and always pay them when scheduled. When their tenure ends, pay them within seventy-two hours of termination, layoff, or quitting. You must document their pay with a receipt or pay stub that includes employee name and Social Security number, pay date and period covered, salary or base wage or piece wage, hours worked or pieces handled, total compensation, state and federal tax deductions plus—with their written permission—other deductions, and net pay.

Employee Well-Being Strategy No. 7: Document!

Make sure you keep good records. Document your employees' hiring, training, and performance as well as your enterprise's performance.

Keep employee personnel files—preferably locked—that include records of exceptional and poor performance. Documentation of verbal warnings and copies of written warnings are vital records if you discharge an employee.

Keep files of workers' compensation claims, discrimination complaints and actions, safety complaints, and Occupational and Safety Health Administration (OSHA) actions. Also place in your files state disability insurance claims, promotion and pay increases, signed resignation forms for voluntary resignation, unemployment insurance claims, vacation and leave requests, medical release to work, plus annual and seasonal pay summaries.

Watch for the new food safety law that was passed in 2011. Called the Food Safety Modernization Act, it will require that you register with federal homeland security if you sell processed foods (such as cheese); depending on your scale, you may be required to develop a hazard analysis and critical control points (HACCP) plan.

Employee Well-Being Strategy No. 8: Provide Safe and Healthy Conditions

You want no one to become injured or ill on your premises. Provide safe and healthy working conditions! For instance, you should provide clean drinking water, spotless restrooms, and

convenient hand-washing facilities. Hang a hand-washing notice in the restrooms. Make sure that all employees attend a job-safety training session as required by California's Injury and Illness Prevention Program. It's a good idea to have one if not all employees take a first-aid and a CPR course.

Even with these precautions, accidents can happen. If an accident does occur:

- get medical treatment for the person
- obtain the injured person's name and address
- call your insurance agent as soon as possible
- document the accident

Employee Well-Being Strategy No. 9: Obtain Workers' Compensation

Workers' compensation provides injured employees specific benefits regardless of who is liable. Obtain workers' compensation; it is required by law. Note that this insurance can be expensive, so factor it in your financial plans.

According to a California agritourism operator in Plumas County, workers' compensation costs him more than liability insurance. It can be complicated, too, with gray areas. For example, if a visitor pays for the privilege of milking a cow for the day, which produces milk that you sell, is that person an employee? (Perhaps it does; contact your lawyer.) Although you might know workers' compensation laws as they relate to your current agriculture operations, new tourism activities can cause confusion.

And they will likely cause changes in your employee classifications and your premiums.

In California, workers' compensation falls under the jurisdiction of the state Department of Industrial Relations. Also under the agency's authority are labor laws, occupational safety and health, and apprenticeship policies. Check out the department's Web site, http://www.dir. ca.gov, and talk with your lawyer to determine your needs.

Special Needs Demand Special Attention

As you consider safety and risk management, consider people with special needs—children, the elderly, and, in accordance with federal law, the disabled. Title III of the Americans with Disabilities Act (ADA) mandates that commercial businesses and private businesses serving the public be designed, constructed, and altered to comply with its specified accessibility standards. These standards involve access to, among other things, lodging, sleeping facilities, restrooms, eating facilities, transportation, and parking spaces.

Information is provided on the ADA Web site, http:// www.usdoj.gov/crt/ada/adahom1.htm. If you have questions, phone the federal Department of Justice ADA information line at 800-514-0301 or consult your lawyer. A good summary of suggestions for complying with ADA is *Entertainment Farming and Agri-Tourism*, by Katherine A. Adam. Figure 5.3 can help you begin to address special needs.

Figure 5.3

Addressing Special Needs

	Yes	No
Building Access		
Do I have parking spaces clearly marked for disabled people?		
Do the spaces comply with the federal Americans with Disabilities Act (ADA) and state law?		
Are the parking spaces near the main entrance?		
Are door entrances wider than 32 inches?		
Are doors easily opened?		
Does door hardware require grasping, twisting, or gripping? (These actions are awkward for elderly, children, or disabled individuals.)		
Do doors have less than 8.5 pounds of pull?		
Do I have revolving doors? (Replace revolving doors with standard doors.)		
Building Corridors		
Are the hallways at least 36 inches wide?		
Are the hallways free of obstacles?		
Is the floor surface hard, level, and not slippery?		
Do obstacles—phones and fountains, for example—protrude into hall corridors and impede passage?		
Do the interior doors have 5 pounds or less of pull?		
Restrooms		
Are the restrooms easily accessible?		
Does door hardware hinder entry? (Handles that require twisting, grasping, gripping, or pinching present difficulties.)		
Are the restrooms large enough for wheelchair-turnaround? (They should be 60 inches minimum.)		
Are stall doors at least 32 inches wide?		
Pathways		
Do pathways have a hard surface?		
Are pathways level or gently sloped, and not slippery?		
Are railings provided in necessary locations?		
Activities		
Do viewing sites allow viewing from a seated or low position, without presenting other dangers?		
Notes		

Remember Your Animals' Welfare

While employee and visitor safety is your priority, animal welfare is crucial. Take good care of your animals. With animal rights a growing concern, it is important you practice defendable health and safety techniques.

There are only a few laws and regulations that mandate animal care practices; these center on animal cruelty and neglect. Many animal care guidelines and numerous publications describe these laws and regulations, such as

- Certified Humane Animal Care Standards, http://www.certifiedhumane.org
- National Safety Council guidelines, http://www.nsc.org/farmsafe/facts.htm#5
- UC Davis School of Veterinary Medicine animal care series, http://www. vetmed.ucdavis.edu/vetext/AN-Progs.html
- Poultry United Egg Producers poultry care handbook, http:// www.fda.gov/ohrms/dockets/dockets/97n0074/c000101.pdf
- euthanasia guidelines (see chapter 7, "Resources for Success," for more information)

Points to Remember

- You are responsible for the health and safety of your visitors and employees.
- While risk and liability cannot be eliminated, they can be reduced and managed by a risk management plan.
- Creating a risk management plan is an essential part of running a successful business.
- There are two components of a risk management plan: farm safety and financial well-being.
- Both components of your risk management plan involve numerous strategies to reduce risk to visitors and employees and, therefore, to reduce your liability.
- You must rely on your management team—particularly your lawyer and insurance agent—to help determine your liability needs.
- You are required by law to address the special needs of disabled people.
- There are a few laws and many guidelines to help you take good care of your animals.

Chapter 6

BLOSSOM TRAIL

EACHES

Designing Your Marketing Strategy

Agritourism in California has the potential to profitably direct market farm products and services, to serve as an alternative use of farm and ranch land, and to supplement your farm income. Creating your marketing strategy and plan of action will help you promote and sell your on-farm products.

What Is a Marketing Strategy?

Your marketing strategy explains how you will promote your agritourism or nature tourism enterprise. It describes what you will offer customers so they walk through your door and what you will do so they come back. It helps you determine who your customers are and how to attract those who most benefit your business. Uniquely your own, your marketing strategy is a function of your products, pricing, promotion, place of sale, customers, competitors, complementary businesses, and your production and marketing costs. Like your business plan, your marketing strategy is fundamental to your enterprise's success. It starts with your business idea and continues through the sale of your product or service. As a result, your marketing strategy is a dynamic process that changes as you evaluate, learn, act, and reflect.

To develop and implement your marketing strategy, begin by reviewing your business plan. Where are you now? Where do you want to be, and how do you get there? Examples and tables throughout this chapter can help you better understand the specific needs and goals of your enterprise. Keep in mind what actions you want to take to attract your customers, encourage them to buy your products, and keep them coming back.

Understand the Market

Agritourism is a great way to add value to your products, and it can help keep you farming. Market the food or fiber you make into a destination. Who lives within thirty miles? With the rise of the local food movement, many of your customers may be within thirty miles of your farm.

Develop Your Brand

We are in a visual age where images—on your Web site and on your various forms of promotional material—speak for your product. Your marketing strategy begins with research. Take time to understand the market in which you'll be working—the world of people looking for entertainment, relaxation, and education on farms and ranches—and the agritourism and nature tourism industry that is ready to offer them just that. Your research will help you evaluate the feasibility of your dreams and uncover information important to your plans.

Know Your Industry

Identify the agritourism and nature tourism trends that can impact your enterprise. Project how the market might change and what to do to keep in step. Are urban "foodies" still excited about eating local food and drinking local wine with famous chefs in orchards? Are U-pick berries popular with large families this year? Did all the other local pumpkin patches add a pony ride or a corn maze? The popularity of social media networking and the Internet mean that social media and a Web site are "must have" promotional tools for your farm or ranch.

You can learn about recent agritourism and nature tourism trends from the following sources:

- topical articles in print and Web-based travel magazines, journals, and newspapers
- free Google alerts for "agritourism" or other keywords that correlate to what you offer
- local agencies such as your visitor's bureau, chamber of commerce, Cooperative Extension office, Farm Bureau, and Small Business Development Center
- Web sites such as the UC Small Farm Program (http://www.sfc.ucdavis.edu/agritourism) and the Agricultural Marketing Resource Center (http://www.agmrc.org/)
- what consumers—your target customers—like and what they avoid

Understand the Customer

Identify your target customers. Discover who is already visiting your area. Tourism boards and your chamber of commerce can provide information about the agritourism or nature tourism market clientele. From this larger market, determine your specific clientele.

Will it be families, teenagers, or people on the go? In 2008, California agritourism operators hosted a wide variety of visitors: families, youth and school groups, individual consumers, wedding parties, reunion groups, artists groups, senior groups, and participants in business retreats. For operators with pumpkin patches and school tours, their visitors were primarily families and younger children. For wineries, U-pick operations, and weddings sites, adults without children were more predominant.

"Selling is getting rid of what you have, while marketing is making sure you have what you can sell," explained one marketer. "The aim of marketing is to know the customers so well that the product fits them and sells itself."

Build Strong Community Relations

Fundamental to any service industry is good public relations. Work to build and maintain a positive image and a sound reputation with your customers, local community, region, state, and industry. Your community can provide valuable emotional, financial, and entrepreneurial support. As you embark on your new venture, become community involved!

Set up a farm FAM tour

A familiarization tour (known as a "FAM tour" in the tourism industry) shows an invited group of participants what a group of agritourism

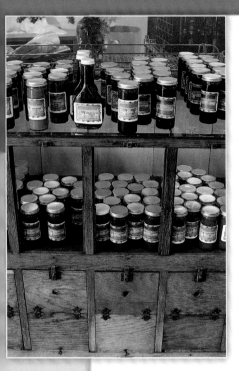

operators in a particular area has to offer. The tour is offered free of charge or at a reduced rate.

You can use the FAM tour as a tool to market your agritourism enterprise directly to consumers. In a FAM tour, you invite potential customers to your farm to view your facilities and learn about its unique activities. If you are planning to host school groups, contact your local schools and invite administrators or teachers out to show them how your activities can benefit or inform their students. Treat them like VIPs.

If your customers are tourists in the area, contact your local chamber of commerce or tourist bureau so they know you are there. Organize a FAM tour for them. You can also invite the media and other operators and community businesses that may complement yours. FAM tour participants are people with the potential to influence others to support or visit the operations on the tour.

Build your off-season offerings

Liberty Hill Farm hosts corporate meetings in its off season in the snowy mountains of Vermont, near Rochester. Cabot Creamery, also in Vermont, hosts meetings at its farm, and afterwards, participants mention Beth and Bob's farm in their blogs. How's that for great advertising?

In Hampshire County, Massachusetts, the Delta Organic Farm focuses on visitors who want to visit and stay at an organic farm, but it also hosts local groups year-round with its conference room and commercial kitchen.

What Makes You Special?

The qualities you offer that make customers feel special are also key to business success. Your unique features distinguish your agritourism or nature

tourism operation from all others. This is also called branding. Identify those features. Do they include any of the following?

- the length of time your operation has been in business (for example, a century-old, family-run farm)
- your location (one hour from the Pacific Ocean)
- the size of your operation (12,000-acre cattle ranch)
- your product or service (outdoor recreation for young singles)
- a unique quality of your product or service (a restaurant on an organic farm)
- benefits of your product or services (solitude)
- unique people involved in your operation (nationally renowned horse trainer)
- your price (affordable family adventure)
- your reputation (featured in "Northwest's Best Places to Stay")
- the lifestyle you offer (the spirit of the West)

Know Your Product

The importance of knowing your product can't be overemphasized. A "product" denotes something that is tangible, designed, manufactured, and packaged. An obvious component of the manufacturing industry, the product is a less obvious element of a service industry. But service industries also have products, and your knowledge of your own is essential to a good marketing plan. When you consider your product, consider your product mix, service, and overall atmosphere or theme. Be authentic.

Product mix

What products will you have on your shelf? Why have you chosen those particular products? For example, if you sell primarily impulse-buy items but carry core items to keep customers coming back, describe your strategy.

Service

When you are shopping, do you prefer hunting the aisles on your own or being assisted by staff? Decide what level your service will be and how it makes sense in your business plan.

Overall atmosphere or theme

What emotions will your customers take away from their experience? Too often, business owners fail to consider how the customer feels after the business exchange. These feelings are critically important to service businesses. In fact, what the customer remembers from the experience is often the only "tangible" product.

Identify Your Features and Benefits

The features of your enterprise are fundamental to its success. Equally important are the benefits that each feature offers. Why? Although the features of your enterprise make it unique, their benefit to the customers draws in clientele.

Tables 6.1 and 6.2 provide examples of features and their benefits. Review them, then write down some features of your enterprise and their benefits to your target customers in figures 6.1 and 6.2.

Develop Your Message

When you know your enterprise's features, you know what makes your enterprise unique—and you can better educate people about your business, both verbally and in writing. In other words, when you know what makes your enterprise unique, you can form key messages for a public relations and advertising program.

With the knowledge of what makes your enterprise unique, you can

Table 6.1

Category of Features			
Product: the definition of your products or services	**Price:** the cost, price, and payment of your products or services	**Promotion:** how you notify your customers of your products or services, and how you sell them	**Placement:** where you sell your products and services
Examples of Features			
Product or service:	**Price:**	**Brochures:**	**Distributors:**
Features: shape, size, package, special characteristics, identification (name, color, logo)	Cost of similar items	Demonstrations	grocery store, farmers' market, restaurant
	Discounts	Samples	Visibility
	Credit terms	Advertising	Ease of purchase
Optional services	Group rates	Sales promotions	Timeliness
Product quality	Weekly rates	Personal sales	Availability
Staff quality		Collaboration	Consumer's awareness of availability
Style		Mailing lists	
Parking		Packaging	Timing
Scenic beauty		Brand or logo	Frequency of service
Location		Location of sale	Tie-in
Guarantees			Co-branding (selling another business's product while it sells yours)
Transportation			
Remoteness			

Marketing Case Study

Gizdich Ranch

When Nita Gizdich first posted highway signs offering U-pick olallieberries, the curious few who trickled onto the farm asked one question: "What are olallieberries?" After Nita changed those signs to read "blackberries" and customers began pouring in, she learned a critical lesson. "Always listen to your customers," Nita says. "They'll tell you exactly what to do."

Since 1970, this marketing philosophy has helped Nita and Vince Gizdich transform their sleepy family farm into a thriving agritourism business. Just outside of Watsonville, California, Gizdich Ranch today attracts suburbanites seeking a taste of the country. Nita, family, and staff offer fresh apple juice, apple pie, apple-themed gifts, and antiques. Year-round, they bake pies and serve deli sandwiches. In April, Gizdich Ranch promotes its apple-blossom splendor. In spring and summer, it welcomes visitors to pick berries and, in fall, to pick apples. What's more, visitors throng to hay rides and thousands of school children swarm to the farm experience.

"We would never have survived just by growing berries and apples," Nita says, noting that only one of the original ten pack-out farms remains in the Watsonville area. So rather than rush the fruit to market, the Gizdichs brought the market to the ranch. And all the while, Nita fiercely promoted their business.

Thirty-five years later, she still campaigns for Gizdich Ranch. She advertises in parenting magazines. She mails postcards to regular customers. Vitally important, declares Nita, is the fact she is always available for the news media: "I don't care *when*. It's *free!*" She even carries brochures while on vacation, distributing them to fellow tourists.

And the highway signs? Nita still posts them, placing the signs on neighbors' land and repaying her friends with all the homemade pie they can eat.

Figure 6.1 YOUR MARKETING STRATEGY

Category of Features

Product:	Price:	Promotion:	Placement:
Describe your products and services	Determine the cost, quality, and payment for your products or services, including your marketing budget	Define your customers and your strategies for reaching and selling to them	Describe your operation outlets for sales and level of service. Compare with your competitors and your marketing allies.

Features Specific to My Enterprise

decide how best to attract customers. You can hire a professional to develop your message or you can brainstorm with family members and outside partners. If you do it yourself, have fun! Be innovative. Remember that your message should attract attention, retain interest, build desire, and encourage a call to action. It should reflect—and be reflected in—your business name, logo, Web site, print materials, and advertisements. When developing your message, ask yourself: what information do I want to provide visitors and what image do I want to project? Pull in the values your family has identified and the unique mix of features and benefits you just uncovered. Then identify your products and services, budget for the enterprise, set prices, determine the method for making reservations, and create clear directions to your site. Once you've determined your message, filter it down to one statement worth remembering and repeating. This makes it easy for others to describe your enterprise.

Launch a Promotional Campaign

Promotion is a big job that requires you to complete tasks in advance and on time. It calls for you to create rates, design and distribute promotional materials, and follow established concrete timelines. Note that publications, seasonal customers, and travel agencies require early notice for their advertising schedules, and community relation-

ships take time and patience to build. So start your promotional activities well before you open your enterprise—three to twelve months ahead of time.

What's more, make sure that every person in your community knows about your new enterprise and what it offers. Word of mouth is the least expensive and one of the most effective forms of promotion. It is also the best way to develop customer loyalty.

Here are some other valuable marketing tools:

- Add a blog to your Web site.
- Put up posters.
- Hand out flyers.
- Collect customers' email addresses and start a monthly e-newsletter with recipes, news about what's fresh, upcoming events, and stories about your animals.
- Distribute brochures and business cards.
- Include recipe cards and bookmarks with products.
- Offer samples, where allowed.
- Provide press releases to local newspapers, radio stations, and television stations.
- Have the local press write a feature story about your unique establishment.

Table 6.2

Sample features and benefits in an agritourism enterprise

Features	Customer Benefits
remote location	rest and relaxation freedom from city bustle clean air unspoiled natural beauty
nearby location	minutes from town oasis in your own backyard family day-trip U-pick farm
farm stand	freshest vegetables experience vine-ripened flavor reminiscent of childhood
small facility	intimate setting exclusive getaway garden cottage fantasy
moderate prices	affordable won't hurt the family budget

Figure 6.2

My Features and Benefits

Features of My Enterprise	Customer Benefits
1.	
2.	
3.	

- Post your media stories on your Web site or Facebook page.
- Tell customers about your product—where it's grown and how it's made.
- Encourage customers to refer you to friends, and offer them a discount for every referral that walks through your door.
- Donate to a local charity or event.
- Work with local restaurants to offer your product on their menu (and make sure your brand name is mentioned).
- Join the local chamber of commerce, or better yet join an agritourism association, if one is nearby.

Whatever marketing tools you select, make sure that they're the most effective ones available for your targeted customers. Don't choose only those you like best or feel most comfortable with. In addition, be consistent with your marketing tools. Don't select promotional methods and then change them before they have a chance to succeed. Too often, a small business owner gets a new idea, modifies the original message or look, and ends up confusing the consumer.

A fictitious example is the 3G Family Orchard with its farm stand and pie shop. Because local surveys indicate most farm-stand customers come from a twenty-mile radius, local awareness was vital to this enterprise's success.

So the operators of 3G Family Orchard posted local road signs. They produced brochures and distributed them at local hotels, motels, tour bus companies, and travel agents. They improved their Web site and designated a Web manager on staff to keep it updated weekly. They also received free local media coverage that stemmed from a newspaper article about the orchard and its history.

The second example was the Working Landscapes Ranch (also fictitious), with its focus on nature tourism. Its operators were targeting young seniors and vacationing families. For promotion, they contacted travel agents specializing in nature tourism, distributed press kits, advertised in an online travel magazine with a sponsored link that fit their customer demographics, and added a YouTube video feature to their Web site that featured a fall foliage roundup.

What Is Your Marketing Cost?

Marketing research, promotion, and continual customer feedback is an ongoing cost of business, so budget for it each year. Your marketing costs depend largely on your enterprise size and type and on your advertising and sales methods. Expect to pay 10 to 25 percent of your total operating costs for marketing during your first four years. As you build a strong reputation and brand, however, your marketing costs will decrease unless competition and other external factors compel you to put more money into marketing to maintain your market share.

Word of mouth was the most common form of promotion of the 332 California agritourism operators surveyed in 2009. Roadside signs, business cards, and brochures, along with a regional guide, were tied with Web sites for the next most popular form of promotion. Feature stories, newsletters, and paid advertising formed the third tier. When asked about the effectiveness of these tools, word of mouth, Web sites, and feature stories rated highest.

Why the Internet Is Essential

An April 2007 survey conducted for Expedia by Harris Interactive asked travelers where they would turn for accurate information for summer travel planning. Online travel sites were the top response (52%), followed by recommendations by family or friends (45%). Rounding out the responses were travel guidebooks (25%), travel community sites (19%), magazines and newspapers (19%), traditional travel agents (17%), and convention and visitor bureaus (16%) (Randall Travel Marketing 2008).

Almost three quarters of California's 98 million travelers made their 2004 travel arrangements online, according to the California Travel and Tourism Commission (2007).

The Internet is widely used every day by members of the general public as their first source of information. The vast majority of California agritourism operators have a Web site; even those spending $500 or less annually on marketing had Web sites. One operator commented, "The Internet is proving to be the biggest PR tool we have. Lots of Bay Area families came after a customer posted a rave review of us."

If you don't yet have any Internet presence, an easy way to start is with a blog on a free site such as WordPress.com or Blogspot.com. You can post a profile of your farm with open hours, directions, and a list of products. You can post and update your events, add photos and YouTube videos, link to your Twitter account, and, per-haps most important, have a Web location where you can direct people for more information and where you can be found by anyone.

Web Sites—Can't Live Without 'Em!

Take a good look at your current Web site and compare it with other agritourism sites you like or have heard about. You can create your own Web site or hire a Web site developer to do it for you. The UC California Agricultural Tourism Directory Web site, http://calagtour.org, has a listing of other California operators, and the North American Farm Direct Marketing Association, http://www.nafdma.com, has some great examples. Regardless of who creates your site, make sure it is easy to use and includes key information such as directions, hours of operation, how to contact you, calendar of upcoming events, products in season now, and customer reviews. Make sure the site is kept current if you want customers to return to it. Keep it clear and simple. Be consistent with information used in your other promotional materials.

What Price?

The price you charge customers reflects what it costs you to manufacture, market, and sell your product or service relative to the features and benefits provided by local competitors. To determine the price, take your break-even point (the cost of business expenses) and add a percentage for profit (your "margin"). If you find yourself charging substantially more than your competitors, review the results of your market research. If you find yourself charging far less, look again at your quality of service; perhaps it needs upgrading.

Consider providing group bookings and large-sales discounts for added profit. Although it's unwise to "buy" business, a smaller margin on a larger volume might earn you money.

Be strategic. For example, consider seasonal prices. If you increase summer prices, you might decrease winter prices too and thus stimulate customer interest during a time you'd otherwise see little activity. Or you might simply save your summer profit for your slow time of year.

Web consultant Gerry McGovern runs a great blog on Web effectiveness at http://www.gerrymcgovern.com/. Here are a few of his tips on effective Web site design.

- Make sure your customers can navigate your site quickly to ensure that you don't lose them, never to return!

- Manage your customers' time. The Web is not free. It charges people for their time. Successful Web sites deliver the most value for the least amount of viewing or navigation time. Google is the benchmark for success on the Web. Google is obsessed with time. Your time. Google is all about helping you find stuff quickly. See what you can do to make your site like theirs.

- Create clear navigation menus for your customers' top tasks and use the words they would search for as they complete the tasks. Good Web navigation is not subtle or overly complicated. It is clear, precise, familiar, and consistent.

Case in point: at the Working Landscapes Ranch's Web site it took less than thirty seconds to find out when its next tour was with the help of a simple calendar on the home page. At the 3G Orchard Web site, the farm's list of upcoming events took two minutes to locate—buried in the last page of the navigation bar titled "Our philosophy."

The more you delete, the more you simplify. The more you simplify, the more you increase the chances of your customers succeeding on your Web site, and the greater the chance they'll return.

Choose the right words. Clear and concise words work best on a Web site. No amount of beautiful images will save you if words can't guide your customers to your information.

How do you rank on Google or other search engines? Having an actively updated (daily or weekly) Web site that has been established for a while places you higher on Google or Yahoo search results. Linking your Web site to other popular and related Web sites will also help new customers learn about you.

Take advantage of online directories. Chileno Valley Ranch uses the Pick Your Own Web site, http://www.pickyourown.org/, to market its fall apple crop and has found it to be very successful in bringing customers to its ranch. This directory, while not fancy, comes up first in searches no matter what terms customers type in, hence its value. See the sidebar for other online marketing directories, some of which are free to join.

Other Web essentials

Domain name: your Web address is important. So is your URL (universal resource locator), which is your online address, so make it short and clear. There are various domain registries online where you register and pay for your name.

Web host: pick a service that will host your Web site. There are many choices, such as Yahoo, Inmotion, iPage, and Wordpress, or you can check Web host sites that list the most popular. Check with another operator to see who they use.

Software: There are many software programs for you or your web designer to use in setting up and maintaining your Web site. Dreamweaver is very popular, but unless you have the time to learn it you may want a Web designer to design and set up your initial site. Another option is Webware, a Web-based design tool that allows you to design and set up your Web site without downloading software to your computer.

Masthead or banner: Look at the mastheads of other Web sites to see what style you like as a potential customer. The Philo Apple Farm in Mendocino (http://www.philoapplefarm.com/), Amy's Farm in Ontario (http://www.amysfarm.com), and Seven Sycamores Ranch (http://www.sevensycamores.com/) in Ivanhoe, near California's Sequoia National Park, are three examples of clear and easy-to-navigate Web sites.

Photos: while some Web sites use lots of photos and few words, people read and use keywords and clear navigation words, not photos, to navigate your site.

Google alerts: Use Google alerts as a way to track your farm in the news and online. Track your free promotions, news stories, and any media source from YouTube to the local press. In Google, type in "google alerts" for an explanation of how this works. Experiment with adding an alert for your farm name and for the word "agritourism." The authors tried this for six months and received on average five to ten alerts per day on various news items from around the nation with some great ideas. If your Web site is not showing up in Google searches, think about what you can do to increase your free media exposure.

Using video and YouTube: Although there are many video-streaming sites, YouTube is by far the most popular. A small, economical hand-held camcorder, such as the Flip or the iPhone 4, allows you to create your own video and post it to YouTube without much fuss in under an hour. From YouTube you can link to your Web site and

Facebook page, and your viewer can easily share your video with others. There are multiple online tutorials to help you create your first video. At last review, there were 468 videos on YouTube for "agritourism." You can add your farm video too.

Think about adding your operation on Google maps at http://maps.google.com/. Type in your farm name or take a look at Work Family Guest Ranch's Web site at http://www.workranch.com to see how much good information can be made available to your potential visitor or guest.

You can also register your farm on Google at http://www.google.com/places and add photos and videos that correlate with other nearby services.

Using Social Media

According to statistics from the California Travel and Tourism Commission (CTTC 2007), we know that

- 86 percent of Americans travel with their cell phones, which they use to call ahead to see what's blooming on the farm today or to book an experience

List Your Business on CalAgTour.org

The UC Small Farm Program hosts a searchable online directory of California agritourism operations for use by visitors looking for a farm or ranch to visit. The directory is located at http://www.CalAgTour.org/.

If you are a working farmer or rancher operating an agritourism business, the Small Farm Program invites you to complete the application online so they can include your business in the directory. If you're already listed, please check your listing and update it if needed. (You can use the sign-up/application form for updates. They will contact you if they have questions.) They are currently updating and planning new promotions for the directory and would love to include more California farms and ranches open to visitors.

Sign up now online!

Online Marketing Directories

UC California Agricultural Tourism Directory, http://www.calagtour.org

Pick Your Own, http://www.pickyourown.org

Agritourism World, http://www.agritourismworld.com

Farmstay U.S., http://www.farmstayus.com

Rural Bounty, http://www.ruralbounty.com

Chefs Collaborative, http://www.chefscollaborative.org

Local Harvest, http://www.localharvest.com

Sleepinthehay, http://www.sleepinthehay.com

- 70 percent of fifteen to thirty year olds use social networks such as Facebook to learn about and share with friends (this usage is growing with older travelers as well)
- 75 percent of Web users trust online reviews more than other written sources

Being visible is paramount. Posted customer comments and ratings are important. Most of all, the visual appearance of your Web presence is crucial, whether it's on your Web site, a Facebook page, your blog, or a Twitter account.

While we know that the Internet is the number one source of travel planning and purchasing, it's the consumer who is becoming the medium or gateway to your farm or ranch via social media and networking sites. The Web site Tripadvisor, which is made up of travelers' reviews, is used by one of four travelers; blogs about your site are also popular sources. Randall Travel Marketing predicts this

Strawberry

Olallieberry

consumer-to-consumer style of travel information sharing will be one of the largest trends to impact the travel and tourism industry in the near future. Simply put, the consumer is now in control of tourism marketing.

If you think Twitter is a type of bird, a blog is a low spot on your farm, and a Facebook page is something you see at the post office, then you need to educate yourself. Plan to attend a regional or national agritourism workshop. The National Farmers Direct Marketing Association is a great resource, as is your local Cooperative Extension, tourist bureau, and resource and development council. In California, there are at least two or three annual workshops about getting started in agritourism.

"The social media revolution is radically changing how direct-marketing farmers communicate with their customers," said Michael Straus, founder of Straus Communications and former vice president of marketing at Straus Family Creamery. "However-

er, it's important to select the right tools for your marketing strategy; otherwise, you could risk a lifetime in Tweeting with insignificant results."

Using Facebook to advertise

Small, niche farm products can be highlighted on Facebook. Use the shop function on Facebook to create a "fanstore" if you plan to ship or sell products by mail.

"[By] using Facebook we are interacting with our customers/fans in a much more direct and immediate manner," comments Michael Zilber, store manager for Cowgirl Creamery. "And from a purely commercial standpoint, we are able to keep them informed on our latest products, specials, and events. But more importantly, we can use it to further our company philosophy and outreach, which helps extend the brand in general. By posting about a variety of subjects related to other cheese makers, artisan cheese in general, and sustainable agriculture, we are furthering content that supports Cowgirl and the issues we think are important to our business."

Use Facebook's reviews wall to post visitor comments. Gather your visitors' e-mail addresses when they come and ask them to sign on as a fan.

Some operators have experimented with online sales via Craigslist. Folks at Rossotti Ranch tried it but reported that "we haven't had much luck with Craigslist. We mainly posted on it hoping we might get a response, but usually don't." They sell most of their meat goats through the Bay Area Meat CSA Web site or to dinner or tour guests to their farm outside of Petaluma in Sonoma County.

Start a Blog

Blogging from your Web site or Facebook is another great way to keep your fans and customers connected to you. Loren Ponica from Stemple Creek Ranch is a daily blogger. Stemple Creek is a family cattle ranch in Marin County, California. The family raises grass-fed beef and lamb on their own organic pastureland just a few miles from the Pacific Ocean. Loren manages the ranch with his father, Al, who, if you asked him about blogging, would probably ask, "Is that a new board game?" Generational preferences count! Remember what we said about current and future visitors and their preferences in chapter 1.

If you don't have a son, daughter, or employee who uses these free promotional tools, look for a volunteer who can help you set these up. Facebook is easy to use and might be the perfect place to start.

So now you're wondering, which one do I set up? The answer is, as many as you can keep up to date! Your Web site and Facebook should sync seamlessly, picking up friends and fans from Facebook and customers via your Web site.

Jane Eckert, a farmer and top agritourism consultant, has many excellent "how to" articles about selecting a Web designer, setting up a blog, creating an e-newsletter for your customers, and more on her Web site, http://www.eckertagrimarketing.com/. She surveyed agritourism operators about their use of social networking and found that 56.5 percent of the respondents were already using Facebook as a marketing tool for their business, and a surprising 65.1 percent were using Facebook for their personal use. Businesses spent an average of one hour per week updating and checking their information, and while some operators used "group" pages as their marketing tool, most used "fan" pages.

Free Social Media Sites

Facebook, http://www.Facebook.com, is a social networking site that connects friends and families. The Web site currently has more than 585 million active users worldwide. Almost half the U.S. population is registered, and there are eight new registrations every second.

Twitter, http://www.twitter.com, is a free information networking and microblogging service that enables its users to send and read messages known as "tweets." Tweets are text-based posts of 140 characters displayed on the author's profile page and delivered to the author's subscribers, who are known as "followers."

YouTube, http://www.youtube.com, is the place to upload videos about your events, testimonials from customers, a virtual tour of what visitors will see at your ranch, and much more. You can then post a link to the video on your Web site, blog, or with Twitter.

Digg, http://www.digg.com, is a social news Web site for people to discover and share content from anywhere on the Internet by submitting links and stories and then voting and commenting on those links and stories.

Stumble Upon, http://www.stumbleupon.com, is an Internet community that allows users to discover and rate Web pages, photos, and videos. It is a personalized recommendation engine that could be very useful for your operation.

Delicious, http://www.delicious.com, is a social bookmarking Web service for storing, sharing, and discovering Web bookmarks.

Reddit, http://www.reddit.com, is a source for what's new and popular online. Users can vote on links that they like or dislike, help decide what's popular, or submit their own links.

TripAdvisor Media Network, http://www.tripadvisor.com, is the largest travel community in the world, with seven million registered members and fifteen million reviews and opinions from travelers.

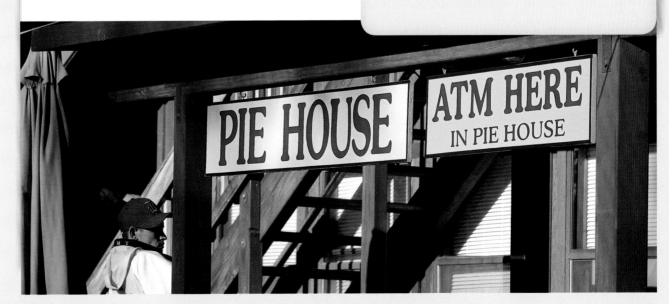

Investigate free promotional Web sites. The best part of these promotional tools is that they are free. The biggest cost is the time it takes for you or someone else to set them up and keep them fresh on a daily or weekly basis.

Check out consumer purchasing trends. The USDA's Farmer Direct Marketing Service, http://www.ams.usda.gov/directmarketing, is a good source. Direct marketing includes farmers' markets, U-pick farms, roadside stands, subscription farming, community-supported agriculture, and catalog sales. The USDA Web

site describes consumer trends in the purchase of fruits, vegetables, nuts, honey, meats, eggs, flowers, plants, herbs, spices, specialty crops, Christmas trees, and value-added products such as cider, jellies, and preserves.

Work with the Press

The media can provide you invaluable exposure and public validation. Develop a working relationship with the media in your area. Using simple and free public relations techniques is one of the most effective ways for your enterprise to get promoted.

Create a Press Kit

Start by creating a press kit, either online or printed. If you are creating a printed press kit, present your material in a folder with sleeve pockets. Either type of press kit should include

- a brief cover letter, including your operation's Web site address and your e-mail address.
- your press release
- two business cards in a printed kit; the same information online
- a brochure, including photos of your farm or ranch
- location and directions
- services provided
- a brief biographical sketch
- press clippings, if available
- testimonials from customers

Actual news is probably the most important element of a press kit. Reporters and broadcast producers receive hundreds of press kits, and—unless you offer them something of news value—yours will likely go into the round file. With an online press kit, you'll be e-mailing your press release to the media. Always answer your phone or e-mail request. Follow up promptly with a courteous reply to make sure the journalist received your release.

Continue to develop a relationship with the local newspaper. You can usually find the e-mail addresses of various reporters at the paper's Web site or at the end of articles. Figure out who covers agriculture and business. Call or e-mail that reporter

From Farming to Facebook: Ten Lessons Learned

1. The world wants to farm. Everybody secretly wants to be a farmer (consider the popularity of Farmville), and Facebook lets you bring your farm to them.

2. Pictures tell a great story. Always carry a camera; take photos with bold shapes and contrasts. Tag your friends.

3. The text should be short and clear. If posts are too long, they will be skipped. "This weekend, our wine club members have an opportunity to pick up their shipments at the Six Sigma tasting room…is too long. Better is: "Come pick up your wine club shipment tomorrow."

4. Small talk wins fans. Fan interaction creates visibility, and visibility wins fans. "It's freezing cold in Asbill Valley this morning. How is your weather?"

5. A few posts per week work well. Too many posts can overload your fans (but posting too infrequently is no good either). Overloaded fans quit following you.

6. Superfans need love. When fans interact with your page often, they should be encouraged.

7. Third-party endorsements are much more valuable than what we say about ourselves.

8. Your fans are your friends. Share content with your fans that you would share with your friends.

9. Keep it fun and positive. Facebook is not the place for criticism or grumpiness.

10. Promote your page! (wink, wink)

Christian Ahlmann
Six Sigma Ranch, Lake County Facebook.com/SixSigmaRanch
http://www.sixsigmaranch.com

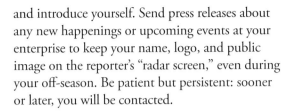

and introduce yourself. Send press releases about any new happenings or upcoming events at your enterprise to keep your name, logo, and public image on the reporter's "radar screen," even during your off-season. Be patient but persistent: sooner or later, you will be contacted.

Find an angle

Consider using one of these angles for your press release:

- strong local story
- public impact
- interesting or unusual information
- useful advice, consumer protection, or other helpful information
- celebrity
- human interest
- timeliness
- proximity
- localizing national trends or stories

Craft a key message

There may be occasions when you need to communicate a specific idea or response to a problem in your operation. Use a public relations tool known as "key messages." Break the information you want to present via the media into three or four main points. These main points are your key messages. Key messages allow you to tell your perspective of the story and provide consistent information to the news media. They help you focus under pressure and may lessen the chance that you'll be misquoted.

Key messages should be the most important information that you want to convey to the public. They should be

- the most essential information, boiled down to a simple sentence
- factual and truthful
- concisely written
- as simple as possible

After you craft your key messages, take a few minutes to familiarize yourself with them. If you have time, rehearse them. Don't memorize them, however, or you'll sound stilted in an interview. Instead, learn each point well and phrase it several ways.

Print journalists call well-composed messages "good quotes," while electronic media call them "sound bites." In a television news story, the average sound bite is twenty-five words or fewer. Note that a sound bite is often used for rhythm and pacing and not for information.

Prepare for an interview

Before you are interviewed, take a moment to anticipate the reporter's questions. If you were the reporter, what would you ask? If you have time, ask your colleagues, partner, or family members this same question. This simple exercise will prepare you for the majority of a reporter's questions.

Also, take a moment before your interview to choose three or four key messages and rate them in order of importance. Your goal is to work your key messages into your answers during your interview. This way you can benefit from your interview by reaching your intended audience with your key messages.

Expect to make one point, possibly two, during an interview. Studies show that most viewers remember just one point of a television or radio show segment.

During the interview

Above all, remain cordial, no matter what the reporter's demeanor. This is an important element of the interview that you can control. If you allow the reporter to upset you, you might lose focus on your key messages and why you agreed to be interviewed in the first place.

When responding to a question, start off by giving a definitive answer—your conclusion—and then explain yourself. This style of answering helps the reporter get quotes and helps organize your thoughts.

While you're being interviewed, imagine that you're talking to an audience, or even a friend, rather than to a reporter. Use conversational language.

Avoid acronyms, jargon, complicated statistics, and technical terms. The point of an interview is to inform your audience members, not to educate them. Therefore, make sure your quotes or sound bites answer the reader's or listener's key question: how does this affect me? Be specific. Keep your answers short, concise, and to the point. Use short examples, illustrations, and anecdotes to illustrate your point. Avoid jokes.

Be truthful. If you don't wish to answer a particular question, reply that you are not prepared to discuss this topic, or it would not be appropriate to comment on that at this time, rather than the evasive "no comment."

Never agree to speak "off the record" and assume what you say next will not appear in print. Follow this simple rule: never say anything you wouldn't want to see in print or on the evening news.

It's helpful in interviews to use bridges. A bridge is a phrase that can lead you from a topic back to your key messages. For example, the time-tested transition "A question I'm often asked is …" can lead you back to a sound bite. Think up a few bridges that are comfortable for you.

Remember that an interview doesn't have to follow a strict question-and-answer formula. Think of it as a conversation with the reporter. It's acceptable for you to sometimes take the lead; if you wait for the interviewer to ask you the "right" question, it might never happen! So answer a question directly and then "bridge" back to one of your key points.

Encourage the reporter to visit your site or to mention it in the story. If you're lucky enough to interest the news organization in a feature story, the reporter will want to come to your farm or ranch. Spend the entire visit with the reporter; provide superb customer service and something for the reporter to take away. Make sure there are customers there the day of the interview who can speak positively about their experience as well.

Target bloggers as another creative way to promote your farm or ranch. Several bloggers write about agritourism and the local food movement. Find them and send them your media packet or a story about you. If they live nearby, invite them out to your site.

After the interview

Be realistic. Don't be disappointed if your story differs from what you expected. Reporters strive for balanced stories, so even what you believe should be a positive story could have one or two negative components.

Once the interview is over and you're back in the office, follow post-interview etiquette:

- Promptly supply any materials you promised to the reporter. Make sure that your contact information is on them.

- Never ask to see a story ahead of time, but always urge the reporter to call you if he or she wants to double-check any of the information that you provided or has follow-up questions.

- Never pressure editors or station managers to prevent a story from running in the newspaper or being broadcast.

- Tell the reporter how much you liked the story. Make yourself available to the reporter for other calls about agriculture or tourism-related topics.

Newsletters

Electronic newsletters, also known as e-newsletters, are another creative way to tell your story and connect with your customers.

As we described earlier, you can blog from your Web site and Facebook page, and you can also send out an electronic newsletter. Some operators have newsletters they mail seasonally, but we encourage you to go paperless. You can get a short, timely newsletter out in a few hours instead of the weeks that it takes to write, design, print, and prepare a bulk paper mailing. You can always print some copies to leave at your counter for customers to pick up, but the e-newsletter is a great way to stay connected to all of your customers. On each e-newsletter be sure to provide your recipients with an opt-out option if they do not wish to receive future copies.

Many online companies offer online e-mail management and other promotional materials such as event fliers, newsletters, and promotions. Constant

Contact, http://www.constantcontact.com, is used by many nonprofit organizations, but there are many others. Take a look at some of the current electronic materials you are getting in your e-mail and scroll down to the bottom of the copy to see who sent it. For a nominal fee, they will manage your list, make flier templates available, and more.

Read the California AgTour Connections' e-newsletter, http://www.sfc.ucdavis.edu/agritourism, about agritourism and nature tourism; it has some great features on agritourism operations, and you can get good ideas on writing tips for your own.

Advertise

Press releases, feature articles, YouTube videos, and media interviews are all free promotion. Advertising, however, is paid promotion. You can pay for time on radio and television, on a search engine like Google (via its online ads and sponsored links), or for space in newspapers, travel magazines, billboards, Web pages, and signs. Although advertising is the most expensive way to publicize your enterprise, it can be a good way to gain first-time clients. Make sure the advertising outlet you choose is the best way to reach your target customers.

Bear in mind, however, that your advertising must be noticeable, and it must be frequent. You should explore the visibility, coverage, timing, and cost of media before choosing one. If you are interested in Internet advertising, Google or Yahoo will have details about how their sponsored links or "adwords" work. Talk to other farms or ranch operators who

have advertised about their success rate. Be sure to advertise only what you can deliver.

At least ten California agritourism operations reported that they spent from $1,000 to $4,999 in 2008 on advertising, yet they did not have a Web site (Rilla et al. 2011).

If we haven't convinced you yet, spend those advertising dollars designing your Web site and get out there. Everyone else is, especially your customers! A Web site can be designed for you for anywhere from free to $2,000. There is no excuse for not having one.

Assess Your Competitive Advantage

Before you launch into advertising, it's important to analyze your competition. By doing so, you'll learn your own strengths and weaknesses and discover some complementary businesses. Collaboration among complementary businesses can increase sales.

Take time to assess your competitive advantage. Identify your competitors, compare your enterprise with theirs, and determine the changes you need for a competitive advantage. Ask yourself these questions:

- What activities can people participate in within my area?
- Do these activities compete with or complement my enterprise?
- Would my targeted customers like to participate in these activities?
  - What other locales—locally, regionally, and in other states—offer experiences similar to mine?
  - What makes my enterprise and locale different from others?
  - How could I collaborate with complementary businesses?

Who is your competition?

Who and what does your product compete with? Who is selling a similar product? Think carefully before you jot down names (use the worksheet in figure 6.3). Your competition could be less than obvious. For instance, television networks consider not only other networks as competition but the

Figure 6.3

Identifying My Competitors

Competitor and Address	How They Compete
1.	
2.	
3.	
4.	

computer- and electronic-game industries too. If you're reserved about looking into other businesses, you might find this exploration uncomfortable—but it must be done!

How do you rate?

Rate your enterprise and the competitors you just identified (use figure 6.4). This exercise can help illuminate changes you can make to gain a competitive advantage.

What could you upgrade?

Based on your notes, write down the upgrades your enterprise might need to improve its competitive advantage (see figure 6.5).

What makes you different?

Consider your distinguishing characteristics once again. What makes your operation special? Write down why your agritourism or nature

Figure 6.4

Comparing My Enterprise with Others

Rate yourself and your competitors on a scale of 1 (poor) to 5 (excellent)

Features (see chapter 2)	My Enterprise Rating	Competitors' Enterprise Rating (numbers keyed to fig. 6.3)			
		1	2	3	4
Image					
Geographic location					
Physical resources					
Accessibility					
Visibility					
Access roads					
Parking					
Nearby transportation					
Nearby lodging					
Products and services					
Customer service					
Advertising					

Figure 6.5
Upgrades for a Competitive Advantage

tourism enterprise stands out above others. You can use the worksheet in figure 6.6.

Sell Smart

Now comes the time to sell your products and services. You need to make your products stand out. You need to make your services attractive. And you must make both your products and services easy to purchase.

Strategize!

- Give your business a personality: choose a name, design a logo, and compose a catchy slogan.
- Present your product with attractive and identifiable packaging: use colors, pictures, and similar packaging for similar products or services.
- Capture the value of your product or service by using phrases like "the spirit of the West" or "a unique weekend getaway."

"Little Things" Count

Promotional items are the extras that add value to your enterprise (see table 6.3). They can increase the appeal of your products and services. Perhaps the most important "little thing" you can offer is availability. Ensure that your product is continually available, unless it is seasonal. If you provide a service, make sure that you are fully staffed and otherwise prepared to offer your service as long as possible. Providing knowledge about your product and your community is important, as is ensuring privacy for your customers if they desire it. If you offer accommodations, support your community by including a small gift in the room that is locally made. A small jar of hand cream or honey is authentic and unique—better than a cheap item made overseas.

Tips for Better Sales

Each agritourism and nature tourism enterprise has its own methods for increasing sales. Incorporate into your selling efforts whatever methods suit your operation. Whatever you do, make sure you charge an adequate fee for your products and

Figure 6.6

What Makes My Enterprise Different?

services. Operators usually charge from $5 to $8 per student for school groups to come to their farm; but many operators we surveyed in California did not charge for this activity in 2008. You provide an essential and unique service and need to be paid for your time and effort. While service activities like tours have a strong marketing angle related to direct sales, other cultural festivals or farm demonstrations are an important source of income. "We have not developed our agritourism into a moneymaking operation," said one operator. "Most visitors are nonpaying customers. We are moving in the direction of paid activities and stays." The sooner the better! See why you need a marketing plan and strategy?

Nature tourism
- Post visible road signs that lead to your operation.
- Know your product and teach your staff about it—whether it's chicken care, pumpkin carving, or canoe paddling.
- Provide as many natural settings as possible.
- Furnish remote and quiet areas that are easy to find.
- Set aside a picturesque spot where guests can be photographed. Offer to take their photograph.
- Erect wildlife feeders and bird nesting boxes to attract a diversity of species.
- Post interpretive signs.
- Provide numerous options for active participation.
- Sell nature goods that are locally produced such as birdhouses, walking sticks, or honey.

- Sell guidebooks, picture books, disposable cameras, film, local postcards, and stamps.
- Hand out brochures and free postcards of your enterprise.
- Post a local map showing areas of special interest.
- Post local nature tourism events.
- List upcoming nature festivals and activities on your Web site.

Table 6.3   Little Things Count

Type of Enterprise	Promotional Items
farm stand	recipes and menu suggestions special packaging theme merchandising processed jams, chutneys contests, coupons, loyalty cards
bed and breakfast	recipes jar of preserves package of cookies
working farm or ranch	locally made hand cream leather keychain T-shirt or hat with logo
hunting and fishing stays	recipes box with tied flies complimentary photograph of guest with catch free shipping home
birding or photo safaris	check-off list of area wildlife complimentary bird checklist
camping	flashlight with logo trail maps wildlife guides point of origin is important; stay close to home

- Develop an electronic mailing list of existing and potential customers via an online marketing service; send them flyers about upcoming events at your operation or in the region.

Shops at farmstays and B&Bs

- Provide baskets with handles.
- Post payment methods you accept such as cash, checks, credit cards.
- Put up signs that identify local goods, with stories and photographs of the maker, farmer, or farm.
- Provide ready-made gift baskets, gift-wrapping, boxing, and shipping.
- Decorate rooms with objects you're selling and place an order pad at the checkout counter.

- Develop and sell a recipe book with recipes that incorporate your products.
- Sell local picture books, disposable cameras, film, local postcards, and books of stamps.
- Set aside a picturesque spot where guests can be photographed; offer to take their photo.

U-pick farms

- Post road signs that lead to your operation and clearly state your business hours.
- Provide different sizes of take-home containers.
- Offer already-harvested crops.
- Post a large map of your operation at the entrance and—if your operation is large—hand out maps to help customers navigate.
- Hire enough staff to provide selection assistance, quick checkout, and carry-out.
- Make sure all staff members wear easily identifiable clothing such as logo T-shirts, hats, vests, and name tags. (This builds team spirit too.)
- Provide hand-washing and restroom facilities.
- Furnish convenient places to sit. The longer customers are on your property, the more they will buy.
- Bring people in one end of the operation and provide parking along the way out. This arrangement brings customers nearer their crop of choice, limits parking problems, and places checkout at the exit.
- Develop a mailing list and a Web site where you offer recipes and announce upcoming crops.
- Post copies of recipes and news of upcoming crops in your booth.

Roadside stands and farmers' markets

- Post easy-to-see prices.
- Have employees wear easily identifiable clothing.
- Provide a choice of products and packaging.
- Sell both prebagged products and bag-your-own products.
- Place baskets and bags in convenient locations.

- Use suggestive merchandising—place complementary products side by side, for instance.
- Offer ready-made gift baskets, gift-wrapping, boxing, and shipping.
- Provide a customer sign-up sheet where you can collect e-mails and mailing addresses
- Post copies of recipes and news of upcoming crops in your booth.
- Know your produce and products from the ground to the table, and teach your staff the same.
- Provide samples, abiding by food-handling regulations.
- Provide good customer service and quick checkout.

Increase Your Off-Season Sales

Even as you breathe a sigh of relief after your last pumpkin leaves the farm or your last Christmas tree is sold, don't forget your off-season! Look around to see what could be offered to customers during your slower time. Develop a conference room in your barn or an outbuilding that can accommodate weddings in the summer and groups in the winter. Cosponsor a community event. Plumas County hosts a wonderful Barns, Birds, and Barbeque Tour that draws birders, history buffs, and businesses. Invite these groups back in the winter to use your conference room. If you have the room and parking, consider cohosting a charity event. Located in rural Sonoma County is the Victorian Christmas Tree Ranch. They cosponsor a very successful Trees for Troops event during their season. Staying involved with your local civic groups gives you the perfect link to these types of events as well as others that could occur during your quieter season.

Evaluate Your Marketing Success

It's important to continually evaluate your marketing success. Evaluation allows you to determine whether you're progressing toward your sales goals and—if you are progressing—whether you're moving as quickly as planned. Evaluation also allows you to deal with problems and identify additional marketing efforts. Customer satisfaction matters. This is the age of the empowered, skeptical, hype-resistant customer. It's a great time to be a customer, and it's a great time to be a customer-centric organization. Customers who are highly satisfied will tell their friends about the good experience. And what drives satisfaction? A good price, a unique experience, and delivery with care to the customer.

One way to be a customer-centric operation is to use a simple customer survey. An informal survey asks customers how they learned of your enterprise, what they enjoyed at your operation, or what you could do to improve it. You'll also want to ask them for their gender, age, size of the group they came with, income range, overall satisfaction, and zip code. Zip codes tell you where your customers live. The more you know your customers and their preferences, the easier it will be for you to customize your efforts to have them coming to your enterprise repeatedly.

You could survey your customers in several ways. Consider placing short and simple comment cards on tables. Have one of your employees gather this information as visitors are wrapping up their visit. You could use volunteers who are fans or friends of your farm, as long as they are a positive, approachable adult or teen.

Online surveys can also be very effective, but you need to take care not to overwhelm your customers with too many questions and remember that completing a survey is optional—a courtesy—not something you can require.

Another evaluation technique is listening for offhand comments. Train your employees to listen when customers wait in line or walk to their car, and ask your employees to write down their comments as quickly as possible. Yet another evaluation technique is performing a head count: simply count the number of people who walk through your door. Are the numbers increasing? Are there peak days of the week or month for visitors?

Coupons, too, can gauge your marketing success. Insert coupons into a mailing and count them when you redeem them. You might even color-code coupons by new versus repeat customers or geographic region.

You might also offer discounts. For instance, you might advertise that if guests mention your radio advertisement they receive a discount of some kind. Keep track of discounts and count them when your promotional period ends. What's more, this method allows businesses with years of information to compare this year's sales data with that of years prior, looking at the same period and taking into consideration weather and disaster-related impacts.

During your regular staff meetings, talk about the requests, comments, and complaints you've received from customers. Listen closely. When your customers ask whether you have something, they're telling you what you should have.

All in all, learn from your evaluation. Creatively use the information you receive. Use it to refine your marketing strategy, hone next year's budget, better publicize and promote your products and services, and—ultimately—increase your sales.

Look After Your Customers

It is important that you serve every person with respect, consideration, and a smile. According to one fourth-generation fruit grower, customer relations are the key to her success. She treats customers as she treats guests in her home so they feel comfortable and want to return.

Welcome!

Your goal is to build relationships that create repeat customers—the best and least expensive marketing tactic. So, start with a good first impression. Personally greet each guest. If you are unable to greet them yourself, have an employee do so.

Introduce yourselves by name and be casual, friendly, and direct. Such a small gesture creates a favorable impression and puts visitors at ease. Keep in mind that first impressions are often derived directly from the person greeting your customers and from the person answering your telephone.

Foresee Customer Questions

In an age when time is the most valuable resource of all, it is vital to answer customer questions as quickly and concisely as possible. This is true on your Web site, and it is also the case for your agritourism experience.

What do you want to know when you visit some-place new? Whatever you want to know is likely to be what your customers want to know. There-fore, prepare yourself with answers in advance.

Begin by anticipating questions. Then make sure that you and your employees can answer these questions. To answer them, your employees must have information (see Alexander 2002). Freely provide it!

Employees must know the history of your enter-prise. They must be able to explain your ideology and practices, including safety measures. When faced with difficult questions, you and your employees need only say that you don't know the answer, but you'll find out—and then follow up! If you can't follow up on the spot, write or e-mail the customer, considering it an added opportu-nity to market your business.

What questions might you be asked? Questions differ, depending on whether your visitors are lo-cal or from afar. Local customers often ask about you and your business:

- How long have you been farming or ranching?

- Why do you farm or ranch?

- What do you do about pests?

- Do you ever let people fish in that stream?

Customers from afar might also ask about the local area and rural living:

- What's it like to live out here?

- How are the schools here?
- What's that bird?
- Is this berry safe to eat?

Customers from afar also might ask about local accommodations and activities:

- Are there any museums or historical sites in the community?
- Where's a good place to stay in the area?
- Can you recommend a good restaurant?
- What activities and events are happening in the next few days?
- Where can I get my car fixed?
- Where will I find tourist information?

Work with Your Neighbors

Good relationships with your neighbors are personally rewarding and professionally important. They are fundamental to your success. Ultimately, good relationships can ward off problems and, with luck, create a spirit of cooperation and collaboration with which your neighborhood can prosper. Think about your neighbors' concerns, perspectives, and values in relation to your own. Whatever your new enterprise, it will impact your neighbors. For

example, what you consider an insignificant farm stand, your neighbors might see as an invitation to strangers and traffic. They might consider hunters as tranquility breakers, bird-watchers as pests, and roads signs as visual blight. Think about how you will address your neighbors' fears and viable concerns. These need attention and resolution (see Kelsey and Abdalla 2008).

What's Your Impact?

If you don't know them already, get to know the people who live near you. Introduce yourself and explain your business, interests, and ideology. Bring them a product sample to demonstrate what your farm generates. Offer them a farm or ranch tour.

It is important to explain your new enterprise early in its development so you can listen to your neighbors' views and accommodate their concerns as much as possible. Be sure to invite them to your open house or grand opening and keep them informed as things develop.

Start your good-neighbor efforts now. Jot down the names of your neighbors, how your enterprise might impact them, and how you can limit negative impact (see the worksheet in figure 6.7).

Figure 6.7

Impact on My Neighbors

Positive Impact	Negative Impact	Methods to Address Negative Impacts

Furthering Neighbor Relations

As you consider the impact of your enterprise, you'll become more aware of your neighbors' concerns. But sensitivity is just the beginning. Good relationships must start well before your first guest arrives.

Think about the following suggestions provided by the Ohio Livestock Coalition on their Web site, http://www.ohiolivestock.org/. This organization works to improve relationships between farmers, ranchers, and neighbors and offers an annual good neighbor award to a nonfarmer. It reminds land-owners that the best public relations technique is one-on-one, face-to-face contact. Furthermore, it says it is important to be a good neighbor—aware, open, considerate, and responsible.

So keep one step ahead of the game. As you work toward your tourism goals, practice being a good neighbor. Try one or more of the following ideas.

- Host an open house or picnic for neighbors during spring.
- Be friendly to neighborhood children. Invite them to see a newborn animal or help them with a science fair project.
- Take opportunities to educate your neighbors about what you do and why. Discuss your enterprise and its specific tasks: for instance, how spreading manure on cropland recycles nutrients and puts the manure to productive use.
- Explain to your neighbors why farmers must work late into the night and on weekends during planting and harvest seasons. If they know the noise, traffic, and lights are limited to certain times of year, they'll likely be more tolerant.
- Spread manure in the most environmentally friendly method.
- Spread manure on any day but Friday—especially not on a Friday before a holiday weekend. Encourage your neighbors to tell you when a fresh dose of manure will infringe on their entertainment plans.
- Help your neighbors. For example, when snow falls, dig them out.

Collaborate

Just as your relationships with neighbors are important, so are your relationships with local businesses. Work with other businesses. You can do this by providing your visitors with brochures

---

Figure 6.8

Businesses I Can Draw On		
Local Business	Supplies/Services They Can Provide Me	Supplies/Services I Can Provide Them

that list other local or related enterprises, which also cultivates your community's economic development (use the worksheet in figure 6.8). You might sell local goods with signs explaining their origin—and have other business owners do the same with your products and literature. Make sure that the businesses you work with share your commitment to quality, however, and monitor their promotional material to guarantee correct use of your brand.

Take a look at Ikeda's Country Market and Perry's Farms in California's Placer County. They provide an example of local collaboration. Ikeda's carries Perry's famous tomatoes, while Perry's carries Ikeda's specialty peaches under Ikeda's name. Each business benefits from the other's recognizable quality. Furthermore, Ikeda's sells also to restaurants that use its brand name on their menu.

Once your enterprise is up and running, work with your competitors (see figure 6.9). No two agritourism or nature tourism enterprises are alike, so refer the clients you can't serve to nearby providers. Find businesses with strengths that complement your own. For example, if you own a B&B and your neighbor offers horseback rides, you might have visitors stay with you and take trail rides with your neighbor. Or if you're a baker and your neighbor is a gifted saleswoman, you might bake bread and have her sell it. You can benefit from collaboration.

In fact, the entire community can benefit from collaboration. One resident's land might be perfect for cross-country skiing while another's is good for hunting and another's is ideal for farmstays. Meanwhile, the local natural area on public land might attract visitors, and your small town might lure urban residents.

Write down what civic groups or chambers of commerce you belong to (see the worksheet in figure 6.10). Think about how you and others can cooperate and collaborate. What businesses in your town can supply your new enterprise? What can you provide them?

From your competitive analysis, who attracts the same kinds of clients you want to attract. How could you work together?

If you offer canoeing opportunities, an outdoor recreation shop and a photography shop complement your enterprise. Who and what local businesses complement your product or service? How might you work with them? What nonprofits and

Figure 6.9	
Competitor	
Competitor	**How We Could Collaborate**

Figure 6.10

Collaborators

Local Businesses, Nonprofits, and Organizations That Complement My Enterprise	Collaboration Methods

Top Ten List of Marketing Ideas for Agritourism Operations

1. Focus your time and energy on having a professional Web site for your business. Absolutely nothing is more important right now than a good Web site. It has now replaced a brochure as your calling card. If you do not create a Web site, you might as well ignore the rest of the marketing ideas on this list.

2. Stay in touch with your best customers through an e-newsletter. It's fast and the most inexpensive way to communicate. Make it easy for customers to sign up with their e-mail address both at your farm and on your Web site. Then make at least monthly contact during the months that you are open. More frequent contact is preferred.

3. Increase your use of promotional offers. All customers today are looking for deals. Use coupons and special offers (such as "Buy 1, Get 1 Free") to entice new and existing customers to return to your business more frequently.

4. Don't rely on the media to find you. Be proactive. Learn your local media and how to write a press release, and get them to do a story on you. A well-placed article about your business is far more effective than any advertising you can purchase.

5. Make a positive first impression! Customers take only a few seconds to form an impression about you and your farm when they drive up to the property. Get rid of the unsightly farm equipment, irrigation piping, chemical barrels, and general trash that detract from a visitor's impression of your farm.

6. Look at the products you sell. The first thing customers always expect is quality. Make sure you listen for customer requests and ask them what else they would like to buy. Be ready to expand your product line to other value-added products, more produce or bakery items, and so on. Customers will give you ideas to increase your sales if you will only ask questions and listen.

7. Focus on your employees. The first step is hiring the right people, but after that make time for training and retraining. Hire people who are willing to learn. Never take for granted that they know your business or how to treat a customer. Also, be sure you set a good example.

8. Connect with your local, regional, and state tourism groups. It is the job of these associations to bring visitors to the area. They have the marketing knowledge and budget to do so. If you don't take advantage of this connection, you are missing out on a tremendous marketing opportunity.

9. The Internet is moving ahead quickly. If you have never heard of Facebook Fan Pages or Twitter, it's time to learn now. While the online social media takes time to develop and maintain, it is a great, FREE resource to spread the word about your business.

10. Get ready to grow beyond your expectations! Most farmers think too small when it comes to making an addition to their building or parking lot. Think big.

Jane Eckert, http://www.eckertagrimarketing.com

organizations might you collaborate with to further business development and community participation? Be sure to continue writing down ways to collaborate as ideas and events develop.

The Central Coast Agritourism Council, http://www.agadventures.org, is a great example of successful collaboration. The council offers marketing exposure that a sole operation couldn't afford or have time to accomplish alone. Groups in North Carolina (http://ncana.blogspot.com) and Hawaii (http://www.hiagtourism.org/) advocate for zoning changes to allow for agritourism. Some agritourism associations offer group insurance. Anything is possible, so think about developing a group in your area that can help with what your enterprise needs. More California agritourism groups can be found in the sidebar.

Joint Marketing

Regional marketing connects people in the agritourism business to work together to promote their industry and geographical area. Apple Hill Growers Association in El Dorado County offers a case in point.

California Agritourism Groups

See the UC California Agricultural Tourism Directory, http://calagtour.org, for the most up-to-date listings.

49er Fruit Trails and Christmas Tree Lane, 530-878-7210, http://www.49erfruittrailandchristmastreelane.com

Apple Hill Growers Association, 530-644-7692, http://www.applehill.com

Calaveras Grown, 209-754-6477, http://www.calaverasgrown.org

Central Coast AgriTourism Council, 805-239-3799, http://www.agadventures.org

Country Crossroads Map (Santa Cruz, Santa Clara, San Benito Counties Farm Trails), 831-724-1356 or 831-688-0748, http://www.sccfb.com/crossroads.htm

El Dorado County Christmas Tree Growers, http://www.chooseandcut.com/

El Dorado County Farm Trails, 530-676-4263, http://www.edc-farmtrails.org

El Dorado Winery Association, 800-306-3956, http://www.eldoradowines.org

Farms of Amador County, 209-223-6482, http://groups.ucanr.org/farmsofamador

Farms of Tuolumne County, 209-928-3775, http://www.farmsoftuolumnecounty.org/map/5

Fresno County Fruit Trail, 559-262-4271, http://www.gofresnocounty.com/Fruit percent20Trail/FruitTrailIndex.asp

Happy Valley Farm Trail (Shasta County), http://www.clearcreekcsd.com/farm.html

Harvest Time in Brentwood, 925-634-4913, http://www.harvest4you.com

Lake County Premium Agriculture, http://lakecountyag.com/

Lake County Farmers' Finest, http://www.lakecountyfarmersfinest.org/

Lodi Wine Trails, http://www.lodiwine.com/winecountry1.shtml

Mariposa Agri-Nature Trail, http://www.mariposaagtour.com/

Mendocino County Promotional Alliance, 707-462-7417, http://www.gomendo.com

Merced County Blossom Trails, 209-385-7403, http://www.mercedrides.com/BIKE/scenic.htm

Oak Glen Apple Growers Association (San Bernardino County), 909-797-6833, http://www.oakglen.net

Placer Grown, 530-889-7398, http://www.placergrown.org/map.php

Russian River Wine Road, 707-433-4335, http://www.wineroad.com

San Mateo County Harvest Guide, 650-726-4485, http://sanmateo.cfbf.com/

Sierra Oro Farm Trail (Butte County), 530-566-9849, http://www.sierraoro.org

Silverado Trail Wineries Association, 707-253-2802, http://www.silveradotrail.com

Sonoma County Farm Trails, 707-571-8288, http://www.farmtrails.org

Sacramento Strawberry Stands, http://cesacramento.ucdavis.edu/Sacramento_Strawberries

Stockton Farms and Wineries, 877-778-6258, http://www.visitstockton.org/agricultural-attractions

Suisun Valley Harvest Trails (Solano County), 707-290-9162, http://suisunvalley.com

Trinity Roots (Trinity County), 530-628-5495, http://www.trinityroots.org/

Yolo County Farm Tours, 530-297-1900, http://www.yolocvb.org/to-do/farm-tours

Yuba/Sutter Agricultural Destinations, 530-743-6501, http://www.visityubasutter.com/agriculture.aspx

Working together

The Apple Hill Growers Association, http://www.applehill.com/, grew out of a small group of orchard ranches that were struggling to survive. Today, it includes more than fifty members—Christmas tree growers, winemakers, and grape growers among them. Its "season" starts in June with a Father's Day cherry festival and runs into December with Christmas tree sales. Apple sales kick off Labor Day weekend. The economic impact of this organization on the county neared $101 million in 2008, according to the El Dorado County Agriculture Department (Shabazian 2010). This is just one example of farmers and the media working together to foster and promote agriculture in their area.

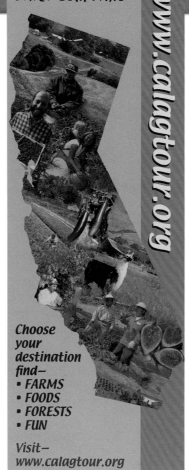

You've Seen Disneyland... Now Visit the Other California

www.calagtour.org

Choose your destination find—
- **FARMS**
- **FOODS**
- **FORESTS**
- **FUN**

Visit—www.calagtour.org

This includes a directory of farmers' markets, roadside stands, pick-your-own operations, farmstays, and other agritourism locations. It also includes information on outdoor recreation opportunities available in Cumberland County. It's a perfect example of joint marketing by a "unified group of private citizens, businesses, not-for-profit organizations, and government agencies."

Learn tourism business skills. Learn the skills required to operate a successful tourism business. Agritourism and nature tourism are service industries requiring an entrepreneurial approach and an understanding of market trends, consumer behavior, consumer attitudes, and consumer preferences. They demand skills and knowledge different from those of traditional agriculture.

Talk to your county economic development director. If agritourism and nature tourism are to become distinct economic development industries, joint promotion, advertising campaigns, and coordinated strategic planning are needed. Cooperation and long-term commitment are necessary for success.

Start your own regional marketing group

The Apple Hill Growers Association illustrates the value of regional marketing for both farmers and local communities. Here are some ideas for starting your own regional marketing program.

All counties have attractions. Some attractions might be agricultural—harvest fairs, demonstration farms, certified farmers' markets, cattle roundups, wineries, microbreweries, and food processing plants, for example. Others might lean toward nature tourism, including natural beauty, national and state parks, monuments, and nature preserves.

Learn what attractions exist in your county and who promotes them. Think beyond simply marketing agriculture. A collaborative effort between the Capital Resource Conservation and Development Council, the Cumberland County Economic Development, and the Cumberland Valley Visitors Bureau in Pennsylvania produced "Local Food, Farms, and Outdoor Attractions."

Approach your county board of supervisors. Join forces with the agritourism or nature tourism operators so you can approach your county board of supervisors as a group. Financial support might be available. For example, money generated by transient occupancy taxes (TOT) has been used in some areas to support and promote regional agricultural marketing efforts, which includes agritourism.

Ask to be included in visitor and economic development endeavors. Use your chamber of commerce membership to ask your visitors' bureaus and county economic development agencies to include agriculture and nature tourism enterprises in destination promotion efforts and materials. There might be visiting groups that would enjoy the experiences you offer.

Participate

As you embark on your agritourism or nature tourism venture, become more involved in your community. Join community organizations and assist with one community improvement project each year. Join the local chamber of commerce to support other businesses and have a voice for your enterprise. Host a chamber of commerce "mixer" at your farm or ranch. Donate products to chamber of commerce endeavors and to other community functions. Operate a "green" business. Recycled products and minimal packaging go a long way toward getting an enthusiastic word-of-mouth review.

Another key way of improving your community relations is spending money locally. When you shop in your community and tell local business-people you're doing so, you'll gain a steady source of referrals. Buying locally can save you money as well. Calculate the cost of vehicle maintenance (see the IRS Web site at http://www.irs.gov for current rates) for your round-trip shopping excursions. Calculate the amount of fuel you use at

Innovative Co-Marketing Groups

Remember the "culture" in agriculture, and check out these innovative tours, festivals, and events that make the most of their seasons and co-marketing efforts. Think about your farm or ranch and how you can tie into larger efforts such as these.

Carson Valley Eagles and Agriculture Tour

Farms of Tuolumne County Farm and Ranch Tour

Flower Fields in Carlsbad

Fresno Blossom Trail

Hoes Down Harvest Festival

Mariposa Agri-Nature trail

Napa Mustard Festival

North Bay Artisan Cheese Festival

Placer Farm and Barn Tour

Placer Mountain Mandarin Festival

Suisun Valley Fun Family Farm Days

Yolo Combines, Bovines, and Fine Wine Tour

today's prices. Then add a wage—even minimum wage—for travel time. The result is the money that you must save each time you shop farther from home.

So, consider adopting a policy of spending some percentage of each dollar at home. What you don't purchase in your hometown, buy as close to home as possible. Use this as an angle on your marketing promotions.

Work with Local Regulators

One more way to strengthen local relations is to work closely with local regulators. During the early planning and development of your business, ask their advice. Not only will this make your work easier, but it also might gain you customers and referrals in the process.

It works! A California rancher who worked with his local legislator wanted to expand his B&B to an overnight lodge. He was frustrated that state statute required overnight lodges to have inspected commercial kitchens (B&Bs need only home-style kitchens). As a result, the rancher contacted his local legislator and convinced her to sponsor legislation allowing full-time agricultural operations to offer overnight stays with meals. From that success came his three-day family farmstays, camp-outs, pack trips, and—soon to be—guest cabins. One rancher's efforts not only changed the law but expanded options for the entire community.

Points to Remember

- A marketing strategy is critical to the success of every agritourism or nature tourism enterprise.

- A marketing plan is what you do to get customers through the door and keep them coming back.

- A marketing strategy has several key components: the market, the enterprise's features and benefits, the message, promotion and advertising, and the competitive advantage.

- Word of mouth is the most powerful and inexpensive promotional method.

- An easy-to-use Web site is an absolute necessity.

- Friendly employees who go the extra mile for your customers are essential.

- Good relations with neighbors, local businesses, and community members are essential to the success of an agritourism or nature tourism enterprise.

- Collaboration with local businesses can be a powerful marketing tool.

- Local residents can work with one another and government agency representatives to begin to revive their agricultural economy.

Acknowledgments

The section "Top Ten List of Marketing Ideas for Agritourism Operations" was provided by Jane Eckert, personal communication. The questions in the "Foresee Customer Questions" section were adapted from P. Alexander and J. Watson-Olson, Starting a Bed and Breakfast in Michigan (Lansing: Michigan State University Extension Bulletin E-2143). The section "Improving Neighbor Relations" was adapted from T. W. Kelsey and C. W. Abdalla, "Furthering Neighbor Relations: Advice and Tips from Farmers" (State College: Pennsylvania State University College of Agricultural Science Web site, 2008).

Chapter 7

Resources for Success

Resources for Success

Help is available for the landowner who wants to become involved in agritourism and nature tourism. Seminars, workshops, publications, and Web sites provide facts, ideas, insight, and advice. This chapter lists some of these resources, particularly those in California, although many national resources are included as well. Keep in mind that Web site addresses change, and you may have to look up key words on your search engine.

General Interest Resources

Directories
Agricultural Resource Directory (California Department of Food and Agriculture)

The *Agricultural Resource Directory* lists agricultural resources, associations, and statistics by crop and county.

California Agriculture Statistics Service
State of California Department of Food and Agriculture
P.O. Box 1258, Sacramento, CA 95812-1258
phone: (916) 498-5161
fax: (916) 498-5186
http://www.nass.usda.gov/ca/homepage.htm

California Agricultural Directory (California Farm Bureau Federation)

The *California Agricultural Directory* lists more than 2,000 agricultural associations and cooperatives, fair associations, and government agencies. It provides key contacts, addresses, telephone and fax numbers, e-mail addresses, and Web sites. This directory is a complete source for farm groups in California, Oregon, and Washington.

California Farm Bureau Federation
2300 River Plaza Drive, Sacramento, CA 95833
phone: (916) 561-5500
fax: (916) 561-5699
http://www.cfbf.com
See also the American Farm Bureau Federation:
http://www.fb.org

Handbooks
Corum, Vance, Marcie Rosenzweig, and Eric Gibson. 2001. *The New Farmers' Market: Farm-fresh Ideas for Producers, Managers, and Communities.* Auburn, CA: New World Publishing.

This book presents in-depth and practical information about selling at farmers' markets for all sizes of farms and markets. The following selections are available as free PDF downloads:

- Educating the Public About Local Agriculture and Farmers' Markets
- Getting Top Dollar for What You Sell at Farmers' Markets
- Top Trends in Farmers' Markets
- Getting Grants for Your Market
- Benefits of Farmers' Markets for Farmers, Customers, and Communities
- Resources from "Embracing the Community" and "Expanding the Vision" chapters.

Gibson, Eric. 1994. *Sell What You Sow! The Grower's Guide to Successful Produce Marketing.* Auburn, CA: New World Publishing.

These books—for farmers and market gardeners interested in alternative marketing options—provide practical information and guidance about selling.

New World Publishing
11543 Quartz Dr. #1
Auburn, CA 95602
phone (toll-free): (888) 281-5170
phone: (530) 823-3886
email: nwpub@nwpub. net
http://www.nwpub.net

Junge, Sharon, Roger Ingram, and Garth Veerkamp, eds. 1995. *Community Supported Agriculture: Making the Connection.* Auburn, CA: Regents of the University of California.

The nonprofit resource Appropriate Technology Transfer for Rural Areas considers this book the best single manual you can buy. It described the book as taking "a potential grower through the process of starting a CSA—including designing the CSA, recruiting members and marketing, creating production and harvest plans, setting share prices, and legal issues."

Small Farm Center. 1994. *Small Farm Handbook.* Davis, CA: Regents of the University of California, University of California, Division of Agriculture and Natural Resources.

This book contemplates all aspects of owning and operating a small farm. Written by the people who know best—farmers, small farm specialists, farm advisors, and researchers—it provides resources, references, and worksheets.

Small Farm Center
University of California
One Shields Ave.
Davis, CA 95616-8699
phone: (530) 752-8136
fax: (530) 752-7716
http://www.sfc.ucdavis.edu/docs/sfhandbook.html
The Small Farm Center's agritourism database can be reached directly from http://www.calagtour.org.

University of California
Cooperative Extension
Placer County
11477 E. Ave.
Auburn, CA 95603
phone: (530) 889-7385
fax: (530) 889-7397
http://ceplacer.ucdavis.edu

Other Resources
Appropriate Technology Transfer for Rural Areas

Appropriate Technology Transfer for Rural Areas (ATTRA) is an important information source for farmers and university extension agents interested in sustainable farming. It is a national sustainable farming information center operated by the private nonprofit National Center for Appropriate Technology. It provides technical assistance, publications, and resources to farmers, extension agents, market gardeners, agricultural researchers, and other agriculture professionals across the United States. The information offered by ATTRA centers on 1) sustainable farming production practices, 2) alternative crop and livestock enterprises, and 3) innovative marketing.

Appropriate Technology Transfer
for Rural Areas
P.O. Box 3657
Fayetteville, AR 72702
phone (toll-free): (800) 346-9140
http://attra.ncat.org/index.html

California Department of Food and Agriculture

This Web site offers news and information about marketing, farmers' markets, food safety, animal health, festivals and fairs, and more.

http://www.cdfa.ca.gov

California Farm Bureau Federation

The California Farm Bureau Federation offers news, information, and benefits about farmer-to-farmer tours, labor and employee services, rural health and safety, and insurance programs. It has 53 county offices in California. Locate the closest office to you from its Web site.

California Farm Bureau Federation
2300 River Plaza Drive
Sacramento, CA 95833
phone: (916) 561-5550
fax: (916) 561-5695
http://www.cfbf.com
http://www.cfbf.com/counties

California Federation of Certified Farmers' Markets

http://farmersmarket.ucdavis.edu

Community Alliance with Family Farmers

The Community Alliance with Family Farmers organizes rural and urban people to foster family-scale agriculture that cares for the land, sustains local economies, and promotes social justice.

http://www.caff.org

National Agricultural Library

The National Agricultural Library is part of the Agricultural Research Service of the U.S. Department of Agriculture. It is a major source for agriculture and related information. Its Web site provides access to its many resources and institutions.

http://www.nal.usda.gov

National Center for Appropriate Technology

The National Center for Appropriate Technology works toward developing healthy communities and an improved quality of life. It provides economically disadvantaged people information and access to technologies. Projects include farmer training, product testing, and demonstrating renewable energy technology.

NCAT Headquarters
P.O. Box 3838, Butte, MT 59702 or
3040 Continental Drive
Butte, MT 59701
phone (toll-free): (800) 275-6228
phone: (406) 494-4572
fax: (406) 494-2905
http://www.ncat.org

News and Information (U.S. Department of Agriculture)

This Web site includes information on farmers' markets, small farms, sustainable development, and food safety.

http://www.usda.gov/news/news.htm

North American Farmers' Direct Marketing Association

The North American Farmers' Direct Marketing Association promotes the growth of farm direct marketing by offering education, networking, and fellowship opportunities. It supplies information on and supports the growth of farm direct marketing. It also voices support for members and the farm direct marketing industry, provides an umbrella organization for regional associations, and encourages the formation of regional and local associations. It hosts an annual conference (and regional conferences) with the most up-to-date info on this topic.

North American Farmers' Direct
Marketing Association
62 White Loaf Road
Southampton, MA 01073
phone: (413) 529-0386
fax: (413) 529-2471
http://www.nafdma.com

Natural Resource Agricultural Engineering Service

This Web site presents numerous publications about farm safety, livestock and poultry, forestry and logging, wildlife, horticulture, and other topics. Many are award winning.

Natural Resource Agricultural
Engineering Service
P.O. Box 4557
Ithaca, NY 14852-4557
phone: (607) 255-7654
fax: (607) 254-8770
http://www.nraes.org/index.html

Office of Small and Disadvantaged Business Utilization (U.S. Department of Agriculture)

http://www.usda.gov/da/smallbus

Small Farms (U.S. Department of Agriculture)

This Web site is a gateway to USDA resources, benefits, and services for farmers.

http://www.usda.gov/oce/smallfarm

Tourism Resources

National Survey on Recreation and the Environment (NSRE)

The 2000 survey represents the continuation of the ongoing National Recreation Survey (NRS) series. Begun in 1960 by the congressionally created Outdoor Recreation Resources Review Commission (ORRRC), the first NRS was a four-season, in-the-home survey of U.S. outdoor recreation participation. Since that time, five additional NRSs have been conducted in 1965, 1970, 1972, 1977, and 1982–83, and one NSRE in 1994–95. Nationwide, more than 50,000 households were included in this survey. Partners include the U.S. Forest Service, the USDA Economic Research Service, the U.S. EPA, and the U.S. Department of Interior's Bureau of Land Management.

http://www.srs.fs.usda.gov/trends/Nsre/nsre2.html

National Survey of Fishing, Hunting, and Wildlife-Associated Recreation

http://fa.r9.fws.gov/surveys/surveys.html

Public Opinions and Attitudes on Outdoor Recreation in California 2002

www.parks.ca.gov
Each state has its own tourism organization or commission with a wealth of information, such as http://www.visitcalifornia.com

New Strategist Publications, Inc.

This organization publishes books about American consumers. It provides information about changing demographics and spending patterns, which can be useful to small business owners.

New Strategist Publications, Inc.
P.O. Box 242
Ithaca, NY 14850
phone (toll-free): (800) 848-0842
fax: (607) 277-5009
http://www.newstrategist.com

Mitchell, Susan. 2002. *American Generations: Who They Are. How They Live. What They Think.* 4th ed. Ithaca, NY: New Strategist Publications, Inc.

This book looks at today's five living generations and at their attitudes, behavior, education, health, households, housing, income, labor force, population, spending, and wealth. Each chapter contains tables and text describing the most important trends and outlook.

http://www.newstrategist.com/bookdetail.cfm/5.htm
http://www.newstrategist.com/productdetails/amgen4contents.pdf

Pamela Lanier's Family Travel

This Web site provides information for families planning a vacation. It furnishes information on lodging, food, adventure vacations, destinations, and deals and specials.

http://www.familytravelguides.com

Travel Industry Association of America

The Travel Industry Association of America supplies information about the U.S. tourism industry and trends that affect it. It provides a list of publications, too.

Travel Industry Association of America
1100 New York Ave., NW, Suite 450
Washington, DC 20005-3934
phone: (202) 408-8422
fax: (202) 408-1255
http://www.tia.org.default.asp

Travelers' Use of the Internet

This TIA report explores the number of U.S. travelers who use the Internet as well as the number who planned or booked online travel this past year. Today, 70 percent of travelers with Internet access use their computer to plan trips.

http://www.tia.org/Pubs/pubs.asp?PublicationID=57

World Tourism Organization

The World Tourism Organization is a leading international organization in travel and tourism. It provides a global forum for tourism policy issues and an information source about tourism.

World Tourism Organization
Calle Capitán Haya 42, 28020
Madrid, Spain
phone: (3491) 567 81 00
fax: (3491) 571 37 33
email: omt@world-tourism.org
http://www.world-tourism.org

California Tourism Resources

California—Find Yourself Here (California Technology, Trade, and Commerce Agency, California Division of Tourism)

This Web site provides tourism information specific to California.

http://www.gocalif.ca.gov/state/tourism/tour_homepage.jsp

Welcome to California (California Technology, Trade, and Commerce Agency)

This state Web site offers information on California business, health and safety regulations, labor and employment, and environment and natural resources.

http://www.ca.gov/state/portal/myca_homepage.jsp

Agritourism Resources

Agricultural Tourism (Small Farm Center, University of California)

This Web site is for landowners interested in agritourism. See the topic "Agritourism and related Web sites," within the Web site. This offers links to sites involved with agritourism, nature tourism, general tourism, tourism and resource sustainability, and value-added agriculture.

http://www.sfc.ucdavis.edu/agritourism/agritour.html

Also on the first page of this Web site is the California Agritourism Database. The California Agritourism Database offers information on California agritourism operations—categorized by region, county, and category of agritourism. It includes such topics as accommodations, direct agricultural sales, educational experiences, outdoor recreation, and entertainment. In addition, it offers an online or downloadable form that California operators can use to get their operations listed for free in the database.

http://calagtour.org

Jane Eckert's firm specializing in marketing agritourism

Eckert AgriMarketing
8054 Teasdale Ave.
St. Louis, MO 63130
phone: (314) 862-6288
fax: (314) 721-0825
email: jane@eckertagrimarketing.com
http://www.eckertagrimarketing.com/agritourism.html

Agritainment: Farm and Ranch Recreation Resource Directory (North Dakota State University Extension Service)

This manual is for farmers and ranchers interested in starting an agritourism enterprise. It offers insights from experienced entrepreneurs about travel trends, tourist needs and wants, tourist activities, breakeven prices, business planning, financing, and more. It provides advice from tax, health, legal, and insurance experts about liability exposure, risks, health regulations, licensing and regulations, sales tax and other tax requirements, risk identification, and insurance coverage.

http://www.ag.ndsu.nodak.edu/ced/resources/farmranch/introduction.htm

Central Coast AgriTourism Council

This is a marketing association of local farmers, ranchers, and tourism professionals promoting their area along California's central coast. You can order a comprehensive map of their area and sites.

http://www.agadventures.org

California Growers Associations

Countless growers associations exist. They range from the California Rare Fruit Growers, Inc., to the California Christmas Tree Association to the California Wool Growers Association. If you want to locate associations pertinent to your enterprise, type "California growers association" into your Internet search engine.

Farm and Ranch Recreation (University of Wyoming)

This Web site contains links to college and university resources, associations, societies, government periodicals, recreation advertisers, and online farm and ranch magazines. The database covers all 50 states. This Web site also offers the *Farm and Ranch Recreation Handbook* by RLS International, which explains how to start and operate a successful farm and ranch recreation business.

http://uwadmnweb.uwyo.edu/RanchRecr

Grudens Schuck, Nancy, Wayne Knoblauch, Judy Green, and Mary Saylor. 1988. *Farming Alternatives: A Guide to Evaluating the Feasibility of New Farm-Based Enterprises.* **Small Farm Series, NRAES-32, Northeast Regional Agricultural Engineering Service Cooperative Extension.**

This guidebook is for rural residents and farmers considering alternative enterprises. Using a format of a case study and workbook, the book helps readers evaluate personal and family considerations, resources, market potential, production feasibility, profitability, cash flow, and all factors combined. It also helps readers build business management skills and offers research sources for enterprise ideas. It provides exercises, self tests, checklists, and worksheets. *Farming Alternatives* was awarded a blue ribbon in the 1989 ASAE (The American Society of Agricultural Engineers) Educational Aids Competition.

http://www.nraes.org/publications/nraes32.html

Alexander, Ben. 2000. *The New Frontiers of Ranching: Business Diversification and Land Stewardship.* **Bozeman, MT: Sonoran Institute.**

This book presents diversification as a way for ranchers to create financially viable and ecologically sustainable operations. It looks at niche marketing, guest ranching, and small business development. More specifically, the book examines the economic potential these opportunities offer to 1) generate income insulated from traditional market fluctuations and 2) provide financial incentives for good stewardship. It emphasizes the relationship between innovative ranch management, enterprise diversification, and a healthy productive landscape.

Sonoran Institute
201 S. Wallace, Box 12
Bozeman, MT 59715
phone: (406) 587-7331
fax: (406) 587-2027
email: sonoran@sonoran.org
http://www.sonoran.org

Great Lakes Coastal Tourism Planning and Development (New York Sea Grant)

This Web site offers publications of interest to agritourism entrepreneurs, including the downloadable publication *Considerations for Agritourism Development* by Diane Kuehn et al. 1998. Oswego, NY: New York Sea Grant, SUNY College.

http://www.cce.cornell.edu/programs/seagrant/tourism

Natural Resources Conservation Service (U.S. Department of Agriculture)

This Web site is devoted to alternative enterprises and agritourism. It includes a "tool kit" with an extensive bibliography, success stories, fact sheets, and more. NRCS co-published the guide "Taking the First Step: Farm and Ranch Alternative Enterprise and Agritourism Resource Evaluation Guide." Call (202) 720-2307 to order or look on their Web site.

http://www.nrcs.usda.gov/technical/RESS/altenterprise/index.html

Farm Trails

Many county and regional organizations across the nation offer marketing and public relations campaigns, educational forums and events, and retail opportunities for their members. In California, these groups include the Sierra foothills' Apple Hill, Grown in Marin, Placer Grown, and Sonoma County Farm Trails. Sonoma County Farm Trails, for example, gears its efforts toward supporting sustainable agricultural diversity in the county. Its Web site features news, events, photographs, membership benefits, and more.

Most California farm trails organizations are listed on the UC Small Farm Center Web site at http://www.catalogtour.org/AgTour.asp?farmtrails=1, with active links to other Web sites.

Tourism Center (University of Minnesota Extension Service)

This Web site has an excellent section devoted to agritourism, which includes publications, materials, and local training opportunities.

http://www.tourism.umn.edu/

Tourism Database (Michigan State University)

This site contains a national extension tourism database of materials related to tourism education. Currently, the database contains hundreds of resource materials including bulletins, reports, videos, and training programs.

http://www.tourism.msu.edu

Nature Tourism Resources

California—Find Yourself Here (California Technology, Trade, and Commerce Agency, California Division of Tourism)

This Web site provides California tourism links.

http://gocalif.ca.gov/state/tourism/tour_homepage.jsp

California Resources Agency

The California Resources Agency oversees the conservation, enhancement, and management of California's natural and cultural resources, including land, water, wildlife, parks, minerals, and historic sites. It is composed of departments, boards, conservancies, commissions, and programs. The Web site links you to these topics as well as to special programs and publications.

http://resources.ca.gov

California Department of Fish and Game

The California Department of Fish and Game oversees California's fish, wildlife, and plants as well as their habitats, managing them for ecological values and for public use and enjoyment. Its Web site offers links to habitat conservation, wildlife conservation, hunting, freshwater fishing, publications, and more.

http://www.dfg.ca.gov/dfghome.html

Ecotourism Development Manual

This manual was a joint project of the NW Arkansas RC&D Council and the University of Arkansas, Cooperative Extension Service.

2301 South University Avenue
Little Rock, AR 72204
phone: (501) 671-2072
fax: (501) 671-2209
http://www.arcommunities.org/tourism/manual.asp

Fermata Inc.

Fermata Inc. works to advance experiential tourism. It inventories nature and cultural resources and develops strategic plans for nature, cultural tourism, and for overcoming tourism challenges. It works with people interested in implementing experiential tourism—organizing marketing plans, developing interpretive materials and training guides, and assessing the impacts of experiential tourism. Its Web site contains information about nature tourism, and for people interested in developing experiential tourism experience in particular.

Fermata Inc.
P.O. Box 5485
Austin, TX 78763-5485
phone: (512) 472-0052
fax: (512) 472-0057
http://www.fermatainc.com

Nature Tourism Information: Welcome to Texas! (Texas Cooperative Extension, Dept. of Recreation, Parks, and Tourism Science, and Texas A&M University)

This site presents information about how to start a business as well as provides answers to questions, examples of enterprises, and an in-depth resource guide. You can also order the handbook "Nature Tourism: A Guidebook to Evaluating Enterprise Opportunities." Topics include options for tourism and recreation businesses, product development, financial plans, marketing plans, legal and regulatory issues, safety procedures, and sources of more information.

http://www.rpts.tamu.edu/tce/nature_tourism

Nature Tourism Planning

This group provides biological and tourism assessments, interpretive and site development, and marketing services for communities, nature tourism businesses, and public sector resource agencies.

1621 13th Street, Suite B
Sacramento, CA 95814
phone: (916) 440-0282
fax: (916) 442-3190
http://www.naturetourismplanning.com

Sustainable Development of Tourism (World Tourism Organization)

This organization provides a global view of tourism and conducts training and conferences on a variety of topics.

http://www.world-tourism.org/frameset/frame_sustainable.html

Watchable Wildlife, Inc.

Watchable Wildlife, Inc., helps local communities realize the economic potential of nature tourism while, at the same time, conserving native plants and animals in their natural habitats. It offers strategies to provide wildlife-viewing experiences, and it works to establish a nationwide network of quality viewing areas. The Watchable Wildlife Web site includes information, publications, and Web site links.

Watchable Wildlife, Incorporated
P.O. Box 319
Marine on St. Croix, MN 55047
phone: (651) 433-4100
fax: (651) 433-4101
http://www.watchablewildlife.org

Business Plan Resources

Web sites offering business plan assistance

Business Owner's Tool Kit

This Web site supplies information to help small-business owners make decisions and "get the most from your business."

http://www.toolkit.cch.com

O'Donnell, Michael. 1991. *Writing Business Plans That Get Results.* Lincolnwood, IL.: NTC/Contemporary Publishing Group.

Many agritourism operators have found the book *Writing Business Plans That Get Results* helpful.

My Own Business, Inc.

The nonprofit organization My Own Business, Inc., offers an online course for starting a new business.

http://myownbusiness.org/course_sba.html

BPlans.com

The private Web site BPlans.com offers a free sample marketing plan for adventure travel.

http://www.bplans.com/spv/3319/index.cfm?affiliate=pas

California Chamber of Commerce

The California Chamber of Commerce helps "businesses do business." Its Web site offers information about starting and maintaining a business in California, including information on labor laws.

http://www.calchamber.com

Canadian Business Service Centers

Important to Americans as well as Canadians, this Web site is a Web-based interactive business planner. It stores your business plan and lets you update it.

http://www.cbsc.org/ibp/home_en.cfm
http://www.cbsc.org/ibp/doc/intro_ibp.cfm

Farm Business Management Information Network for British Columbia, FBMInet-BC

Also applicable to the American entrepreneur, this site contains integrated financial statements and enterprise budgets.

http://www.farmcentre.com/english/checkup/index.htm

Microsoft Office

From version '98 on, the full version of Microsoft Office contains an interactive business plan writer that incorporates users' financial statements. It is available in the "Templates" folder.

The Business Plan: Road Map to Success (U.S. Small Business Administration)

This Web site is an online business plan tutorial, in English and Spanish.

http://www.sba.gov/starting_business/planning/basic.html

The U.S. Small Business Administration's Small Business Start-Up Kit

A paper copy of this kit is available through your regional Small Business Administration office or local Small Business Development Centers (SBDC). Or you can download this and other useful forms from the SBA Web site.

http://www.sba.gov/starting/indexstartup.html

Virtual Business Plan

By the developers of Business PlanPro software, this site offers good quick advice.

http://www.bizplanit.com/vplan.htm

Local Organizations Offering Business Plan Assistance

Small Business Development Centers (SBDCs) provide management assistance to current and prospective small-business owners. SBDCs offer individuals and small businesses a wide variety of information and guidance in easily accessible branch locations. The program is a cooperative effort of the private sector, the educational community, and federal, state, and local governments. Your local SBDC is an important place to visit. Its National Information Clearinghouse Web site lists a wealth of information.

http://sbdcnet.utsa.edu/default.htm

U.S. Small Business Administration

The U.S. Small Business Administration (SBA) offers valuable Web resources and personal help to small-business owners. It provides direct and guarantee loan programs, and it cosponsors the Small Business Development Centers that exist in conjunction with California's Community Colleges and Economic Development agencies. There are six regional offices in California, all listed on the SBA Web site. See the SBA Web site for additional information.

http://www.sba.gov

Other Local Resources

Other helpful resources for starting your tourism enterprise are your local convention and visitor bureaus. All are active in attracting visitors, and some are active in attracting agritourists and nature tourists specifically. They are administered through county government and funded through the Transit Occupational Tax (TOT). Ask your county government for their numbers—and, at the same time, ask about local economic development corporations and economic development districts. Most counties have local economic development boards and commissions composed of helpful staff. Also check out your chamber of commerce.

Financial Resources

Many organizations help small businesses obtain funding. You should look into the following organizations and contact your county government as well. Most counties have an economic development person who knows lenders in your county.

The Farm Credit System

The Farm Credit System is a network of borrower-owned lending institutions and related service organizations that offer financial help to farmers across the United States and Puerto Rico. These institutions provide credit and related services to farmers, ranchers, and aquatic-product producers and harvesters, and they issue loans for processing and marketing. In addition, the institutions issue loans to rural homeowners, certain farm-related businesses, and agricultural, aquatic, and public-utility cooperatives. The Farm Credit System recently started pilot programs for beginning farmers, young farmers, and part-time farmers. For the contact in your area and for further information, look up its Web sites.

American AgCredit, ACA
200 Concourse Blvd.
Santa Rosa, CA 95403-1120
http://www.agloan.com

http://www.fca.gov
http://www.fccouncil.com

USDA Farm Service Agency

In most counties, the Farm Service Agency is located within the office of the U.S. Department of Agriculture. The Farm Service Agency lends money and provides advice and credit counseling to farmers. It advises farmers and ranchers temporarily unable to obtain private commercial credit such as beginning farmers who can't qualify for conventional loans because of insufficient net worth or established ranchers suffering financial setbacks from natural disasters. It issues loans to farmers and ranchers for purchasing farmland and financing agricultural production. The Farm Service Agency has two basic loan programs—the Guaranteed Loan Program and the Direct Loan Program. See its Web site for details.

http://www.fsa.usda.gov

USDA Rural Development

Also within your local consolidated USDA office is the Rural Development department. Rural Development offers a Business and Industry Guaranteed Loan Program that guarantees as much as 80 percent of a loan issued by a commercial lender to start or expand a rural business or cooperative. This loan program reduces the risks faced by private lenders and raises their lending limits. It might help applicants otherwise unable to obtain credit for a business start-up.

Rural Development offers grants as well. Its grants include those to economic and community development agencies, which pool funds for micro loans and issue them to start-up businesses and small businesses that cannot otherwise obtain funding. Its Web site offers further information, including the addresses and phone numbers of regional Rural Development offices.

http://www.rurdev.usda.gov

Small Business Financial Development Corporations

Small Business Financial Development Corporations are nonprofit public-benefit corporations chartered and regulated by the Office of Small Business within the California Technology, Trade, and Commerce Agency. There are eight FDCs in California. Through them, smal-business owners might obtain a California Guarantee Loan.

The California Loan Guarantee Program seeks to create and retain jobs and to provide service to small businesses. It provides lenders the security—the "guarantee"—they need to approve a loan or line of credit. It thereby enables small-business owners to obtain a term loan or line of credit for which they otherwise cannot qualify. Moreover, the program allows these businesses to establish favorable credit with a lender, which is important to obtaining additional loans without FDC help.

For additional information about the California Loan Guarantee Program, look at its Web site.

http://www.caloanguarantee.ca.gov

Community Development Financial Institutions

Community Development Financial Institutions provide a range of financial products and services. Such products and services include commercial loans and investments to start or expand small businesses and funds for low-income households and local businesses. Community Development Financial Institutions also provide technical assistance to small businesses and credit counseling to consumers. Through these institutions, farmers and ranchers can apply for financing from the Community Development Financial Institutions Fund. This fund was created in the U.S. Department of Treasury to expand the availability of credit, investment capital, and financial services in distressed urban and rural communities. Details are furnished on the Community Development Financial Institutions Fund Web site.

http://www.ustreas.gov/cdfi/

Marketing Resources

Ag Marketing Resource Center, Iowa State

This center, located at Iowa State, provides a national information resource for "value-added agriculture" for farmers and processors. Several other land-grant universities are partnering on this USDA-funded project.

Ag Marketing Resource Center
1111 NSRIC, Iowa State University
Ames, IA 50011-3310
phone (toll-free): (866) 277-5567
fax: (515) 294-9496
email: AgMRC@iastate.edu
http://www.agmrc.org

Agricultural Marketing Service (U.S. Department of Agriculture)

www.ams.usda.gov/directmarketing

Direct Marketing Series (Appropriate Technology Transfer for Rural Areas)

www.attra.org/attra-pub/directmkt.html

King, R. 2000. *Collaborative Marketing, A Roadmap & Resource Guide for Farmers.* University of Minnesota Extension Service Bulletin BU-07539-GO.

http://www.extension.umn.edu/distribution/
businessmanagement/DF7539.htm

Valdes, Isabel. 2000. *Marketing to American Latinos: A Guide to the In-Culture Approach.* **Ithaca, NY: Paramount Market Publishing, Inc.**

This publication offers facts and figures to help understand the size and power of the rapidly growing U.S. Hispanic market—an increasingly important market for agritourism and nature tourism. It introduces the "New Latina" and "Generation N," and advises how to market to them.

phone (toll-free): (888) 787-8100
http://www.paramountbooks.com

Management Resources

Farm Management Publications (Farm Management Resources, Washington State University)

This Web site offers numerous inexpensive publications about both farm management and alternative crops plus some interesting Web links.

http://farm.mngt.wsu.edu/misc.html

Savory, Allen. 1999. *Holistic Management: A New Framework for Decision Making.* **Covelo, CA: Island Press.**

http://www.holisticmanagement.org

Legal Resources

California County Agriculture Commissioners (California Department of Food and Agriculture, Welcome to California)

The names and contact information for all local agriculture commissioners are listed on the California Department of Food and Agriculture Web site as a PDF download.

http://www.cdfa.ca.gov

Copeland, John C. 1998. *Recreational Access to Private Lands: Liability Problems and Solutions.* **2d ed. Fayetteville, AR: National Center for Agricultural Law, Research, and Information, University of Arkansas, School of Law.**

This book is available from the University of Arkansas's National Center for Agricultural Law, Research, and Information.

University of Arkansas School of Law
National Center for Agricultural Law
Fayetteville, AR 72701
phone: (479) 575-7646
http://www.nationalaglawcenter.org/index.html

Employment Development Department Forms and Publications (California Technology, Trade, and Commerce Agency, Welcome to California)

This Web site provides access to publications about disability insurance, work opportunity tax credit, employment tax, unemployment insurance, recruitment and referral services, and more.

http://www.edd.ca.gov/formpub.htm

Hamilton, Neil D. 1999. *The Legal Guide for Direct Farm Marketing.* **Des Moines, IA: Drake University.**

This book provides information to help farmers, USDA employees, and other advisors understand the effect of various laws and regulations on direct farm marketing. It provides general information and advice on how the law might pertain to specific situations, and it addresses liability insurance questions.

Drake University Agricultural Law Center
2507 University Avenue
Des Moines, IA 50311
phone: (515) 271-2947
http://www.iowafoodpolicy.org/legalguide.htm

National Center for Agricultural Law Research and Information

The National Center for Agricultural Law Research and Information (NCALRI) addresses legal issues affecting American agriculture. These issues include government farm programs, farm finance and credit, hired labor practices, land use, liability and insurance coverage, environmental law, international trade, and biotechnology. The NCALRI conducts research and analysis and provides up-to-date information to farmers and agribusinesses, attorneys, community groups, and others confronting agricultural law issues. It focuses on research, writing, publishing, developing its library services, and disseminating information to the public. It does not provide legal representation or advice about specific legal questions.

http://www.nationalaglawcenter.org

Understanding Farmers Comprehensive Personal Liability Policy: A Guide for Farmers, Attorneys and Insurance Agents. **Fayetteville, AR: National Center for Agricultural Law, Research, and Information, University of Arkansas, School of Law.**

This book discusses interpretation of the common farm liability policy. Its friendly, question-and-answer format makes for easy reading. It is available from its publishers.

University of Arkansas School of Law
National Center for Agricultural Law
Fayetteville, AR 72701
phone: (479) 575-7646
http://www.nationalaglawcenter.org/index.html

Safety and Risk Management Resources

Agricultural Health and Safety Center at UC Davis (Centers for Disease Control and Prevention)

The UC Davis Agricultural Health and Safety Center is one of nine agricultural health and safety centers established in the United States by the Centers for Disease Control and Prevention. Its mission is to protect and improve the health and safety of U.S. farmers, farmworkers, and consumers. Its Web site describes the center's news and events, current research, publications, and useful links.

http://agcenter.ucdavis.edu

Agricultural Safety and Health (National Institute for Occupational Safety and Health, Centers for Disease Control and Prevention)

The National Institute for Occupational Safety and Health is the federal agency responsible for conducting research and making recommendations about preventing work-related disease and injury. Its Web site supplies links to other sites plus information about agricultural safety and health.

http://www.cdc.gov/niosh/topics/agriculture

NAFDMA Liability Insurance

Due to changes in the insurance industry, North American Farmers Direct Marketing Association (NAFDMA) no longer offers a liability insurance plan. However, the association does offer its members access to a referral list of approximately 20 insurance companies that provide insurance to farm direct marketing operations.

http://www.nafdma.com/Public

A Safer Site: Agricultural Safety and Health (National Safety Council's Youth Activities Division)

This Web site is designed for kids of all ages, parents, teachers, and youth group advisors. In addition to fact sheets, it provides links to related Web sites and organizations.

http://www.nsc.org/mem/youth/9_top.htm

Appropriate Technology Transfer for Rural Areas

This Web site provides information about such topics as protecting against liability and guarding against risks to children.

http://attra.ncat.org/attra-pub/entertainment/other.html

Facts about Agricultural Safety and Health (National Institute for Occupational Safety and Health, Centers for Disease Control and Prevention)

This Web site provides interesting facts about safety and health on the farm and ranch.
http://www.cdc.gov/niosh/agfc.html

FoodSafe Program (University of California at Davis)

This Web site examines "hot topics" such as foot and mouth disease and looks at the food industry, food industry referrals, and pesticides. It also provides links to other resources dealing with food-safety issues.

http://foodsafe.ucdavis.edu

Injury and Illness Prevention Program (University of California at Santa Cruz, Environmental Health and Safety)

This Web site offers information about avoiding various injuries and illnesses.

http://ehs.ucsc.edu/injury_illness_prevention

National Children's Center for Rural and Agricultural Health and Safety

The National Children's Center for Rural and Agricultural Health and Safety works to improve the health and safety of children exposed to the hazards of farming, ranching, and rural living. Of special interest to the agritourism and nature tourism operator is information about the National Adolescent Farmworker Occupational Health and Safety Advisory Program and North American guidelines for children's agricultural tasks. This Web site also provides links to other sites.

http://research.marshfieldclinic.org/children

National Safety Council

The National Safety Council's Agricultural Division presents fact sheets on farm safety, from livestock handling to sun and heat exposure.

http://www.nsc.org/farmsafe/facts.htm

Safety and Health Resource Guide for Small Businesses (National Institute for Occupational Safety and Health, Centers for Disease Control and Prevention)

This Web site provides information about small-business safety and health resources and links to other Web sites, too.

http://www.cdc.gov/inosh/topics/smbus

Safety, Health, and Environmental Resources (National Safety Council)

The Safety, Health, and Environmental Resources Web site provides information important to workplace safety, health, and environmental resources.

http://www.nsc.org/index.htm

Workers' Compensation Site (California Department of Industrial Relations)

The Department of Industrial Relations was established to improve working conditions for California's wage earners and to advance opportunities for profitable employment in California. Its Web site contains information about workers' compensation.

http://www.dir.ca.gov/workers'_comp.html

Resources for Addressing Special Needs

Americans with Disabilities Act (U.S. Department of Justice)

This Department of Justice Web site provides information on the Americans with Disabilities Act, including the Act itself. It includes a description of technical programs provided to small businesses, enforcement of the Act, certification, mediation, plus new and proposed regulations. It also includes important links to relevant Web sites.

http://www.usdoj.gov/crt/ada/adahom1.htm

Appropriate Technology Transfer for Rural Areas

This Web site contains information about complying with the Americans with Disabilities Act.

http://attra.ncat.org/attra-pub/entertainment/other.html

Hunting and Fishing Access for Disabled People (California Department of Fish and Game)

This Web site discusses programs that offer hunting and fishing opportunities for disabled people.

http://www.dfg.ca.gov/coned/access.html

Animal Health and Welfare Resources

Animal and Plant Health Inspection Services (U.S. Department of Agriculture Marketing and Regulatory Programs)

This Web site provides current information about animal diseases and suspected outbreaks.

http://www.aphis.usda.gov

Animal Health Branch (California Department of Food and Agriculture)

This Web site offers a series of important publications to inform and educate people about diseases and disease prevention.

http://www.cdfa.ca.gov/ahfss/ah

Dr. Temple Grandin's Web Page: Livestock Behavior, Design of Facilities and Humane Slaughter

This Web site discusses modern methods of livestock handling that improve animal welfare and productivity. It offers information on the behavior and handling of cattle, pigs, antelope, bison, and other animals. It also contains information on the design of animal-handling facilities, humane slaughter, and animal welfare.

http://www.grandin.com

State Humane Association of California. 2002. *California Animal Laws Handbook*. Sacramento, CA: State Humane Association of California.

This publication lists animal laws, including those about cruelty. In it are some laws relating to farm animals. It is updated annually.

P.O. Box 2098
El Cerrito, CA 94530
phone: (510) 525-2744
http://www.californiastatehumane.org

United Egg Producers

The United Egg Producers publishes a valuable animal-husbandry document. Check out its Web site.

phone: (770) 587-5871
www.fda.gov/ohrms/dockets/dockets/97n0074/c000101.pdf

University of California, Davis, Veterinary Medicine Extension

This Web site provides news, information about California programs, and online publications about animal health and production. More specifically, it includes information about livestock and poultry diseases. It offers information on the ecology of diseases and food safety as well as outreach information and Web site links. There is an "animal welfare program" page, and eight publications that address the issue of animal care relating to food-animal production in California.

http://www.vetmed.ucdavis.edu

County Planning Resources

California Planners Information Network (California Office of Planning and Research)

This Web site furnishes information about California's local planning agencies. It discusses an annual survey that identifies recent planning activities, accomplishments, and trends, and it provides information and results. It also supplies the names and addresses of key planning officials, the current status of local general plans, and other information about specific jurisdictions.

http://www.calpin.ca.gov

The California Planners' Book of Lists **(California Office of Planning and Research)**

This Web site contains contact information for California city and county planning agencies. It provides a directory of local planning agencies called the *California Planners' Book of Lists.* The Web site is updated annually.

The State Clearinghouse
phone: (916) 445-0613
http://www.calpin.ca.gov/Archives/Default.asp

Niche Market Resources

Postharvest Research and Information Center (University of California)

The UC Postharvest Research and Information Center works to 1) improve the quality and value of horticulture crops available to the consumer, 2) reduce postharvest losses and improve marketing efficiency, and 3) solve particular problems in handling fruits, vegetables, and ornamentals to maintain their quality and safety. Its Web site furnishes information about its courses, workshops, activities, and publications. It provides information about postharvest studies, online postharvest data, and a postharvest resources directory. It also provides links to useful Web sites.

http://postharvest.ucdavis.edu

Vegetable Research and Information Center (University of California's Division of Agriculture and Natural Resources)

The Vegetable Research and Information Center collects and distributes information important to consumers, growers, and processors of California's vegetable industry. It conducts research and provides information in support of the industry. Its Web site includes information about vegetables, issues, news and events, useful links, and much more.

http://vric.ucdavis.edu

Conservation Resources

California Association of Resource Conservation Districts

The California Association of Resource Conservation Districts is a voluntary association that helps California Resource Conservation Districts meet conservation goals. It promotes federal, state, county, and municipal agency cooperation with the districts. There are now 103 California Resource Conservation Districts, most of which are funded primarily through grants and some of which are funded by county property tax revenues. They also receive training, in-kind support, and a watershed grant program from the Department of Conservation and the Natural Resource Conservation Service.

The California Association of Resource Conservation Districts Web site lists all Resource Conservation Districts. You may have to search on the Web site to find the one nearest you; district names are not necessarily linked to counties. For

example, Sierra Valley Resource Conservation District and Feather River Resource Conservation District both are located in Plumas County. Search within "Find Your RCD" column.

http://www.carcd.org
http://www.nrcs.usda.gov/partners/districts.html (National Resource Conservation District Web site)

Duryea, Mary, ed. 1988. *Alternative Enterprises for Your Forest Land.* **University of Florida Extension, Extension Bulletin #810.**

http://www.sfrc.ufl.edu/Extension/pubtxt/cir810.htm

Farmland and Open Space Resources (California Department of Conservation's Division of Land Resource Protection)

The Department of Conservation's Division of Land Resource Protection provides information about conserving California's farmland and open spaces. Working with landowners, local governments, and researchers, it offers programs to guide land use planning decisions that help farmers and ranchers protect their land.

http://www.conservation.ca.gov/dlrp

Farmland Trusts

Conservation easements protect farmland from urbanization and allow farmers to continue farming. As a result, they are attracting interest. Explore the following Web sites for more information:

http://www.farmland.org (American Farmland Trust)
http://www.rangelandtrust.org (California Rangeland Trust)
http://www.lta.org (Land Trust Alliance)

Natural Resources Conservation Service

The Natural Resources Conservation Service—formerly the Soil Conservation Service—is a federal agency in the U.S. Department of Agriculture. It works closely with landowners to conserve natural resources on private lands. The Natural Resources Conservation Service Web site has a wealth of information about conservation programs, technical help, and the location of the nearest service center.

http://www.ca.nrcs.usda.gov

Educational Resources

A Farmer's Guide to Hosting Farm Visits For Children. 1998. San Francisco, CA: The Center for Urban Education about Sustainable Agriculture and Market Cooking for Kids.

The goals of this booklet are to share the good educational ideas already being practiced on farms and to encourage and facilitate farm field trips. It provides workable, effective ideas for planning and hosting educationally powerful visits. It offers activities for elementary school students, many of which can be adapted to other age groups. Log onto the UC Sustainable Agricultural Research and Education Web site to review this guidebook.

http://www.sarep.ucdavis.edu
http://sarep.ucdavis.edu/grants/reports/kraus/97-36FarmersGuide.htm

Agriculture in the Classroom (U.S. Department of Agriculture)

This Web site offers an agricultural curriculum and activities. It provides teacher resources, links to state programs, children's activities, and information on the Agriculture in the Classroom National Conference.

http://www.agclassroom.org

California Agricultural Core Curriculum Advanced Clusters

This Web site presents lesson plans about agricultural topics geared for high school and community college students, but they can be adapted to younger children as well. They provide activity ideas for agritourism and nature tourism operators as well.

http://www.calaged.org/ResourceFiles/Curriculum/AdvCluster

California Foundation for Agriculture in the Classroom

This Web site supplies resource materials, teacher training opportunities, student programs, kids corner, and other links.

www.cfaitc.org

Pizza Farm

The innovative Pizza Farm was developed by Darren Schmall at the Madera County Fairgrounds. Darren found that the most effective method of reaching and teaching children about agriculture was through a favorite food— and he chose pizza. His Pizza Farm illustrates the potential for farm tours. Its Web site offers further information.

www.pizzafarm.org
http://www.pizzafarmok.org

Appendix A

Farm Visits for Children

The most common agritourism activity is hosting school field trips. Operators in California charge from $5 to $10 per child, depending on the length and depth of the experience. Make sure you charge a fee. What you do is important, and you need to be compensated for the value you provide.

At Harley Farms, for example, the owner Dee Harley celebrates the arrival of baby goats by welcoming school groups and other visitors into her world for a few hours. She gives anywhere from two to ten tours per week, year-round. The tours cost $10 per person and comprise 25 percent of annual revenue. "We couldn't survive on tours alone, but the money we make from them certainly allows us to stay small and dedicated to quality artisan cheese-making," and she adds, "it's become an irreplaceable part of our income" (*New York Times*, February 17, 2007).

The Farm, in Salinas, is an agricultural education center, demonstration farm, and produce stand that is designed to tell the story of contemporary farming along California's Central Coast. It offers a wide variety of primarily organic fresh fruits, vegetables, and fresh-cut flowers and host tours for all types of groups, including school groups, businesses, senior organizations, study groups, and travel tours. The Farm tailors its program to best suit the age level and interests of each group. The school or group tours also include a hay wagon ride, the opportunity to harvest a crop, and an introduction to farm animals. In 2010, the fees for the basic one-hour school or group tour was $8 for adults, $6 for children under sixteen, and children two and under free with a $25 minimum charge for the group. (Teachers and chaperones of school groups were free). This agritourism operation has really refined their pricing to fit the size and type of group.

Agritourism operators have a great opportunity to educate children. Most children who visit your farm or ranch will have limited or no exposure to agriculture and know little about where their food comes from. You can expose these young visitors to agriculture and provide them firsthand experience with growing and raising food. Sharing your farm with children safely and effectively requires some know-how and advance planning.

Share and Teach

If you know about growing and harvesting plants, for example, you can talk about planting, irrigating, weeding, harvesting, packing, and selling. You could emphasize that different crops grow in different seasons: orchard fruit grows in the spring and summer; broccoli and spinach grow into late fall and winter. You can explain or show when different tasks occur: planting in spring, harvesting in summer and fall, and maintenance in winter. Take time to plan your themes, topics, and message.

Provide Hands-On Activities

People learn most easily with hands-on experience. When children experience the message they hear, they relate better to the message, have more fun, and retain more of the information. Increase the hands-on activities in your farm tour or event, and you'll likely enhance your visitors' experience.

Your farm or ranch teems with opportunities for hands-on activities. For example, you might have children plant seeds, transplant greenhouse sprouts, weed gardens, harvest crops, taste vegetables, pack fruit, make compost, pet animals, or feed animals. Brainstorm! List the activities you perform every month and identify those that children might enjoy. As you work with more and more groups, you'll come up with more and more ideas.

Keep in mind that hands-on activities require planning time, set-up time, flexibility, and a sense of humor. Give yourself ample time to prepare. And when you're conducting these activities, relax and have fun. Remember: this is not about a perfect end product but rather an enjoyable farm or ranch experience for your young visitors.

Different Ages, Different Activities

Certain activities work best for certain age groups. For example, clearing big weeds in an orchard

is fulfilling for young children, while weeding garden beds is rewarding to older children. You must understand some basic developmental characteristics of children to effectively teach them.

First, consider young children. They have a short attention span: a five-year-old's is no more than fifteen minutes. They find it hard to follow written or verbal instructions. They learn by practice. And they are interested in doing the activity, not necessarily in what it produces. Therefore, your activities with young children should include visuals in a step-by-step process, plus demonstrations. Short activities are always better than long ones.

Now, think about older children. Their attention span is at least thirty minutes. They care about the product as much as the process. In other words, the activity you offer older kids should have a product at the end. If you offer a corn maze, provide older children with a certificate if they find the middle or make stops along the way where they gather and record information for a prize.

Once you identify some hands-on activities appropriate to specific age groups, test them. Try them on children you know, on your own kids or family friends. In this way, you can determine how well the activities work, where problems may arise, the amount of time they require, and what supplies you need. Solicit feedback from the children to bolster your capability, confidence, and success.

When you conduct a hands-on activity with a large group of kids, break up the group. Select a group leader for each subgroup. These leaders can

Exploring More On-Farm Educational Opportunities

A lavender farm has each child create a lavender wand to take home; during harvest, each child picks a pumpkin to carry away. A dairy farm shows their groups how to make butter in a churn, then they taste it with on-farm honey viewed through a clear-sided demonstration bee hive.

Besides public school groups, many private, charter, and home school groups are looking for unique field trips. Get a list of these schools from your local school district or county office of education or find them online. In addition to children and teen groups, college-level students make great visitors. Check with your local community college agriculture or natural resource departments. There is nothing like a real "field research" experience. Interns are also a source of assistance. Many young people today are looking to learn skills on farms. An agritourism operation could be the perfect place for this exchange. The Worldwide Opportunities on Organic Farms, http://www.wwoofusa.org/, is one way to connect with young people. You might want to consider using teens or college students who are in a recreation or teaching program to conduct your tours if you are too busy with your farm stand. The possibilities are endless.

Farmers and ranchers operating education programs might also find some useful resources and connections through the Farm-Based Education Association, an organization that strengthens and supports the work of educators and administrators who provide access to their productive working farms. Although members are primarily from the eastern half of the United States, many of the ideas shared are universal.

Appendix A, Table 1.   A checklist for planning for groups.

Planning Item	Information to Communicate or Gather
Weather	Make sure the group leader knows what kind of weather to expect and that he or she plans accordingly. Have a "rain policy" in case of bad weather, either a back-up activity or postponement.
Traffic and parking	Post signs and place cones to direct traffic in and out of your operation. Make sure your parking area is identified and accessible to buses. (See chapters 2 and 4)
Group size	Determine the largest group size that your operation can accommodate at one time. This often is related to the number of people in your business or family who can lead a group of children. (See chapter 2.)
Number of groups in a day	If you're working with several schools, there might be several groups wanting to visit in one day; decide whether you want to host groups for the entire day or for just part of it. You must determine the amount of time your daily chores require.
Age of children	Find out if your visitors comprise younger children, older children, or a mix. This will help you select appropriate activities. Give a nametag to each child.
Chaperone number and role	Determine a minimum and maximum number of chaperones. For example, during hands-on activities, you might break a group of 24 students into three groups of 8— meaning you have a minimum of 3 and a maximum of 6 chaperones. Make sure chaperones understand their roles, which might include keeping the group together, helping group members perform activities, and managing the group.
Tour length, travel directions, travel time	Identify and communicate the length of time of your tour. Provide detailed directions to your operation, and help determine travel time to and from it. Discuss your policy regarding latecomers and no-shows. (See chapter 2.)
Special interests	Explore ways to develop a theme that integrates the group's special interests, school studies, or reasons for visiting your farm or ranch.
Food	Make plans for eating. Provide water. Perhaps group members will bring their own lunch or snack—but if you plan to supply food at an additional price, check local regulations. (See chapter 4.)
Restrooms and washing hands	Provide restroom and hand-washing facilities. Do you need portable toilets? (See chapters 2 and 4.)
First aid and emergencies	Find out if there are special needs or health concerns in the visiting group. Have an easily accessible first-aid kit. Make sure that someone on your farm has CPR training. Develop plans for emergencies, from bee stings to broken arms. (See Chapter Four.)
Insurance	Get liability insurance. Make sure the group leader signs a waiver. Find out what insurance policy the group carries. Always monitor safety on your operation, before and during the tour.

check supplies, answer questions, help others, and maintain order.

Making Arrangements

There are many considerations to address before a visiting group arrives. First is cost. Most farm tours run from $5 to $10 per person. It's smart to ask other farm-tour operators what they offer and what they charge.

Second, prepare for your visitors. Spend time before with their teacher or group leader, explaining your requirements and learning the group's needs. You might suggest pre-tour and post-tour activities or reading materials that the group leader can share with the class or group.

Some ideas to help you prepare for visiting groups can be found in table A.1. It is important you read chapter 5, "Developing Your Risk-Management Plan," as well.

Touring the Farm

Once the group has settled, tell the children who you are, the history of your farm, and what you do. Ask who has visited a farm or ranch and what they did or saw. Encourage the kids to ask questions, to participate in activities, and—importantly—to have fun!

Show your visitors the restroom and hand-washing facilities. Explain ground rules such as the following:

- Be respectful of your home. This farm or ranch is your home, and they should treat it as such, leaving it as they found it and picking up trash before they go.
- Obey safety rules. Identify what these are.
- Stay away from off-limit locations. Identify what and where these are.
- Stay on designated paths and with group leaders. Move quickly to the tour stops.
- Respect all animals. Show the children how to interact with animals and instruct them to wash their hands after they pet them.

Next, jump into an activity. Pursue a hands-on activity that captures everybody's attention. You want to involve the children in a fun and interesting experience, yet leave them time for play and fun. Be sure to give them something to take home, such as a small gift or a sample of what they have harvested. This will help them remember and possibly return with their family.

At the end of the visit, summarize what the children have seen and experienced. If you ask them what they enjoyed most and what they liked least, you'll learn what you need to improve. Finally, of course, thank the kids for coming!

Farm Tours for Profit

A recent UC survey of California agritourism operators (Rilla et al. 2011) found that even though more than half of respondents conduct farm tours, most do not charge a fee for the tours. Philip McGrath, owner of McGrath Family Farm in Ventura, and Helene Marshall, of Marshall's Honey in Marin, have figured out how to operate farm tour programs that contribute to their businesses' bottom lines and do not take too much valuable time.

Three years ago, four thousand students from Los Angeles Unified School District visited McGrath Family Farm, financed by grants. In 2009, Philip McGrath hosted groups from schools in Watts every month and was waiting for word about funding from the CDFA Specialty Crops Program to pay for more student field trips to the farm. He suggests that farmers who would like to host school field trips contact school districts directly to learn about such programs.

Helene Marshall has been showing curious people around her beekeeping and natural, gourmet honey production operation since she started selling at farmers' markets, but soon found out that this took valuable time away from other things that needed attention. The Marshalls now do tours by reservation only and on limited days. Sunday tours, offered twice a month for $25 per adult or $10 for kids, are two hours long and include a twenty-minute film produced by the National Honey Board, a honey and food pairing, tastes of up to twenty-five varieties of honey, a look at an opened beehive, a taste right out of the beehive, and a demonstration of extracting and bottling the honey. School group tours are available on Wednesdays and Thursdays for a minimum fee of $250.

Marshall says, "Farmers often forget to put a value on our time, and that's the one thing that we don't get back." Although tours of Marshall's Honey operations are not a huge moneymaker, they are never a loss.

Penny Leff, excerpted from AgTour Connections Newsletter September 2009.

Appendix B

Planning a Tourism Workshop

Many of you already know how to plan and conduct workshops. But do you know how to organize a session oriented specifically toward agritourism or nature tourism? The following tips can help. The more sessions you prepare, the more experience and contacts you'll gain in this field—and the more knowledgeable you and your community will be.

General tips

Look for partners

It could be that some natural resource agencies and agricultural organizations are thinking about the same topics you're planning to present. Find out, and work with them if so.

Can your Small Business Development Center help?

Your local Small Business Development Center is a great resource for business planning. Ask if it will provide your participants with individual follow-up business consulting.

Visit the Small Farm Program Web site

Log onto the UC Small Farm Program Web site to locate materials (http://www.sfc.ucdavis.edu). Check out its agritourism database to find local presenters, panel members, and participants.

Organizing the workshop

Present the facts

Be realistic about the potential that agritourism and nature tourism offer. Remember that you provide the information and your participants decide what to do with it and how.

Helpful materials for your workshop

- Copies of *Agritourism and Nature Tourism in California*
- Handouts from speakers and other local people
- Sample of local promotional and entrepreneurial materials, including business cards and brochures
- A bulletin board or table on which participants can place their own materials
- Audio-visual equipment
- Flip charts and marking pens
- Workshop evaluation forms

Look at past workshops

In appendix B, figure 1, you will find a sample workshop agenda for an agritourism and nature tourism program. Use it for ideas (but do not draw from its list of speakers; it is vital you cultivate and harvest local talent!).

Holding the workshop

Use this manual

Agritourism and Nature Tourism in California is written for you! It is both a reference guide and a training guide—so when you customize your training to meet workshop goals, feel free to refer to individual sections. Review the "Chapter Goals" and "Points to Remember," and decide which chapters are most useful. If you hand out a complete manual, explain which sections you're skipping during the session and why.

Ask participants to come prepared

Depending on your goals and anticipated outcomes, suggest that registered participants arrive with a completed homework assignment. Many chapters of this manual are based on assessment. So ask participants to complete certain worksheets *before* the session and to bring them to the session. Then, build on these worksheets during the workshop.

Present real-life experience

Include speakers involved in the agritourism or nature tourism business. Have them give presentations, provide demonstrations, and take part in panel discussions. Participants of previous agritourism and nature tourism workshops often report this to be the most useful aspect of the class.

Ask your participants

Use your participants as resources. In one conference room, you'll have an abundance of shared knowledge.

Provide names and numbers

As part of your handout materials include a participant roster so people can contact each other afterwards.

Encourage mixing

Include an introductory icebreaker that encourages participants to meet and be comfortable with one another.

Vary learning styles

Break up the learning style. Mix visual presentations with group exercises and other training styles so that participants stay interested and engaged.

Accommodate your audience

Check in with your audience. Do they look bored or are they actively involved? Accommodate them accordingly.

Take a trip!

Consider an afternoon field trip to an agritourism or nature tourism enterprise.

Wrapping up the workshop

Acknowledge speakers and people who helped.

Have participants complete a workshop evaluation form, which will supply feedback for your next workshop. Gather the evaluation forms from participants, compile them, and give them to your organizing team. Appendix B, figure 2, shows a sample evaluation form.

Appendix B, Figure 1

Short Course: Agricultural and Nature Tourism
Program Agenda
Morning Session (9 a.m.–12:30 p.m.)

9:00–9:20 **Introduction/Overview of the Day**

Shermain Hardesty, Director, UC Small Farm Program

9:20–9:55 **The Market for Agriculture and Nature Tourism**

Jonelle Tannahill, Rural Tourism Liaison, California Travel and Tourism Commission

9:55–10:15 **What Is It? Is It for Me?**

Ellie Rilla, Community Development Advisor, UC Cooperative Extension, Marin County

10:15–10:30 **Assessing Your Assets**

Ellie Rilla, Community Development Advisor, UC Cooperative Extension, Marin County

10:30–10:55 **Alternative Ranching Enterprises**

Holly George, Livestock and Natural Resources Advisor, Plumas-Sierra Counties

10:55–11:30 **Developing a Business Plan & Raising Capital**

Small Business Development Center

11:30–12:30 **Getting Permits and Clearances**

Moderator: *Penny Leff*, Small Farm Center

Jrennifer Barrett, Planner, Permit and Resource Management Department, Sonoma County

Katherine Kelly, Impossible Acres Farm

Elaine and George Work, Work Ranch

Lunch **Box lunch provided**

Afternoon Session (1:30–4 p.m.)

1:30–2:30 **Product Development, Marketing, Advertising and Promotion**

Moderator: *Holly George*, Livestock and Natural Resources Advisor, Plumas-Sierra Counties

Darren Schmall, Pizza Farm

Katherine Kelly, Impossible Acres Farm

Karen Bates, The Apple Farm

2:30–3:00 **Risk Management**

Peter Krause, Allied Insurance Services

3:00–3:30 **Consumer Relations and Employee Development and Management**

Nita Gizdich, Gizdich Ranch

3:30–4:00 **Wrap Up/Closure**

Penny Leff, Director, Agritourism Coordinator

Appendix B, Figure 2

Agritourism and Nature Tourism Workshop

Evaluation Form

Please answer each question as honestly as possible. Your comments will help us evaluate this workshop and plan future events.

Overall, how would you rate this seminar?

☐ Excellent   ☐ Good   ☐ Satisfactory   ☐ Poor

Comments: _____

_____

_____

Did it meet your expectations? Why or why not? _____

_____

_____

How can we improve this seminar or others? _____

_____

_____

What future topics or programs would you find interesting? _____

_____

_____

The following is the list of presented topics. Please rate them as

4=Excellent   3=Good   2=Satisfactory   1=Poor

The Market for Agricultural and Nature Tourism	4	3	2	1	N/A
What Is It? Is It For Me	4	3	2	1	N/A
Assessing Your Assets	4	3	2	1	N/A
Alternative Ranching Enterprises	4	3	2	1	N/A
Developing a Business Plan and Raising Capital	4	3	2	1	N/A
Getting Permits and Clearances	4	3	2	1	N/A
Product Development, Marketing, and Advertising	4	3	2	1	N/A
Risk Management	4	3	2	1	N/A
Consumer Relations and Employee Development	4	3	2	1	N/A

Thank you for your help.

References

Adam, K. L. 2004. Entertainment farming and agri-tourism. National Sustainable Agriculture Information Service Web site, http://attra.ncat.org/attra-pub/entertainment.html.

Agri-Business Council of Oregon. 2007. Agri-tourism workbook. ABC Web site, http://www.aglink.org/agbook/agritourismworkbook.php.

Alexander, P., and J. Watson-Olson. 2002. Starting a bed and breakfast in Michigan. Lansing: Michigan State University Extension Bulletin E-2143.

Beus, C. E. 2008. Agritourism: Cultivating tourists on the farm. Pullman: Washington State University Extension Farming the Northwest Publication EB2020.

Brown, D. M., and R. J. Reeder. 2007. Farm-based recreation, a statistical profile. USDA Economic Research Report 53.

California Watchable Wildlife. 2010. Web site, http://www.cawatchablewildlife.org/.

CTTC (California Travel and Tourism Commission). 2007. Rural tourism strategic plan (2007–13.) CTTC Web site, http://www.visitcalifornia.com/media/uploads/files/00698CTTCRuralStrategicPlan_3.pdf.

———. 2008. Data tables. CTTC Web site, http://tourism.visitcalifornia.com/media/uploads/files/editor/Research/2008.

Egan, T. 1998. Indian reservations bank on authenticity to draw tourists. New York Times (September 21).

ERS (U.S. Department of Agriculture Economic Research Service). 2009. Data set for median household income. ERS Web site, http://www.ers.usda.gov/data/unemployment/RDList2.asp?ST=CA.

Friemuth, J. 2001. Farming experience attracts tourism. Capital Press Agriculture Weekly (August 31).

Garrison, B., et al. 2005. Development and marketing strategies for birding and wildlife tourism in the greater Reno, Nevada region. Nature Tourism Planning Web site, http://www.naturetourismplanning.com/pdfs/Marketing%20Plan%20%20Final%20%203.16.05.pdf.

Geisler, M. 2011. Agritourism profile. Ames: Iowa State University Agricultural Marketing and Resource Center. AgMRC Web site, http://www.agmrc.org/commodities__products/agritourism/agritourism_profile.cfm.

George, H. 2008. Agritourism enterprises on your farm or ranch: Where to start. Oakland: University of California Agriculture and Natural Resources Publication 8334. ANR CS Web site, http://anrcatalog.ucdavis.edu/Items/8334.aspx.

Gibson, E. 1994. Sell what you sow! The grower's guide to successful produce marketing. Auburn, CA: New World Publishing.

Hamilton, N. D. 1999. The legal guide for direct farm marketing. Des Moines, IA: Drake University.

Hunter, M. 2006. A growing taste for culinary travel. CNN Travel, September 1. CNN Web site, http://www.cnn.com.

ILO (United Nations International Labour Organization). 2010. Tourism can be a "major generator" of jobs after economic crisis. UN News Service (November 19) Web site, http://www.un.org/apps/news/story.asp?NewsID=36813&Cr=touris&Cr1=&Kw1=UN+international+labour+organization&Kw2=travel+and+tourism&Kw3=.

Kelsey, T. W., and C. W. Abdalla. 2008. Good neighbor relations: Advice and tips from farmers. State College: Pennsylvania State University College of Agricultural Science Web site, http://pubs.cas.psu.edu/freepubs/pdfs/ua309.pdf.

Klotz, J. 2002. How to direct-market farm products on the Internet. Washington, DC: USDA Agricultural Marketing Service. USDA Web site, http://www.usda.gov/wps/portal/usdahome.

Krippendorf, J. 1986. The new tourist: Turning point for leisure and travel. Tourism Management 7 (June): 131–135.

Leff, P. 2011. California counties adapt permitting and regulations for agritourism. California Agriculture 65(2): 65.

Leonard, J. 2008. Wildlife watching in the U.S.: The economic impacts on national and state economies in 2006. Addendum to the 2006 National survey of fishing, hunting, and wildlife-associated recreation. Arlington, VA: U.S. Fish and Wildlife Service. U.S. Census Web site, http:nat.survey 2006_economics.pdf.

Lopinto, N. 2009. Der Indianer: Why do 40,000 Germans spend their weekends dressed as Native Americans? Utne Reader Web site, http://www.utne.com/Spirituality/Germans-weekends-Native-Americans-Indian-Culture.aspx.

National Geographic Society. Center for Sustainable Destinations Web site, http://travel.nationalgeographic. com/travel/sustainable/.

O'Donnell, M. 1991. Writing business plans that get results: A step-by-step guide. Chicago: Contemporary Books.

Ohio Livestock Coalition. 2011. It takes two. Ohio Livestock Web site, http://www.ohiolivestock.org/ Publications.html.

OTTI (U.S. Department of Commerce Office of Travel and Tourism Industries). 2010. 2008 U.S. travel and tourism statistics. OTTI Web site, http://tinet.ita.doc.gov/outreachpages/outbound.general_information. outbound_overview.html

Outdoor Industry Foundation. 2007. The next generation of outdoor participants. Outdoor Industry Foundation Web site, http://www.outdoorindustry.org/images/researchfiles/ResearchNextGeneration. pdf?55.

Pollan, Michael. 2006. The omnivore's dilemma. New York: Penguin Press.

Randall Travel Marketing. 2008. Top ten travel and tourism trends for 2007–2008. Mooresville, NC: RTM. North Carolina State University Extension Web site, http://www.ncsu.edu/tourismextension/resources/ documents/top10traveltrends.pdf.

Rilla, E., S. D. Hardesty, C. Getz, and H. George. 2011. California agritourism operations and their economic potential are growing. California Agriculture 65(2): 57–65.

Rosenzweig, M. A. 1999. Market farm forms: Spreadsheet templates for planning and organizing information on diversified market farms. Auburn, CA: Full Circle Organic Farm.

———. 2001. Management skills for agriculturalists. Davis, CA: Farm and Agriculture Collaborative Training Systems (FACTS).

Shabazian, D. 2010. Connections: Sacramento region urban rural connections strategy. March newsletter. SACOG Web site, http://www.sacot.org/rucs.

Smart, R. B. 2009. Agritourism: Make your farm a destination. Every Kitchen Table blog (October 23). http:// everytable.wordpress.com/.

Smith, B. 2011. Senate passes bill limiting agritourism liability. Indiana Public Media. 8 March. Indiana Public Media News Web site, http://indianapublicmedia.org/news/senate-passes-bill-limiting-agritourism- liability/.

Space Coast Living. 2011. Birds of a feather: Birding and wildlife festival. Space Coast Living (January 6). SCL Web site, http://www.sclmagazine.com/birds-of-a-feather-birding-and-wildlife-festival/.

TIES (The International Ecotourism Society). 2006. Global tourism factsheet. TIES Web site, http://www. ecotourism.org.

Tosetti, C. 2001. Agritourism adds value to farming. Capital Press Agriculture Weekly (August 31).

TTIA (The Travel Industry Association of America). 2004. Domestic travel market report. Washington, DC: TTIA.

———. 2007. The ideal American vacation trip report. Washington, DC: TTIA.

U.S. Census Bureau. 2010. S2501 Occupancy characteristics. Data set from 2005–2009 American community survey. U.S. Census Web site, http://factfinder.census.gov/servlet/STTable?_bm=y&-geo_id=01000US&- qr_name=ACS_2009_5YR_G00_S2501&-ds_name=ACS_2009_5YR_G00.

U.S. Forest Service. 2003. National survey on recreation and the environment (NRSE). USFS Web site, http:// www.srs.fs.usda.gov/trends/Nsre/nsre2.html.

U.S. Small Business Administration. 1991. Small business start-up information package and regional resource guide. Spokane, WA: SBA.

USFWS (U.S. Fish and Wildlife Service). 2000. The bottom line: How healthy bird populations contribute to a healthy economy. USFWS Web site, http://library.fws.gov/Pubs/mbd_bottom_line2.pdf.

Ward, J. L. 1997. Growing the family business: Special challenges and best practices. Family Business Review 10(4): 323–337.

Wine Institute. 2006. Report on the economic impact of California Wine. St. Helena, CA: MKF Research.

Wood, M. E. 2002. Ecotourism: Practices and policies for sustainability. Paris: UN Environment Program.

Index

Page numbers in **bold type** indicate major discussions. Page numbers in *italic type* indicate tables.